Titles in Series:

High Performance Liquid Chromatography

Analytical Chemistry by Open Learning

Second Edition

Author:
SANDIE LINDSAY
Newham Community College

Editor:
JOHN BARNES

Published on behalf of Thames Polytechnic, London
by
JOHN WILEY & SONS
Chichester · New York · Brisbane · Toronto · Singapore

Copyright © 1987, 1992 Thames Polytechnic, London, UK

Published in 1992 by John Wiley & Sons Ltd
                      Baffins Lane, Chichester
                      West Sussex PO19 1UD, England

All rights reserved.

No part of this book may be reproduced by any means,
or transmitted, or translated into a machine language
without the written permission of the publisher.

Other Wiley Editorial Offices

John Wiley & Sons, Inc., 605 Third Avenue,
New York, NY 10158-0012, USA

Jacaranda Wiley Ltd, G.P.O. Box 859, Brisbane,
Queensland 4001, Australia

John Wiley & Sons (Canada) Ltd, 5353 Dundas Street West, Fourth Floor,
Etobicoke, Ontario M9B 6HB, Canada

John Wiley & Sons (SEA) Pte Ltd, 37 Jalan Pemimpin #05-04,
Block B, Union Industrial Building, Singapore 2057

Library of Congress Cataloging-in-Publication Data:

Lindsay, Sandie.
    High performance liquid chromatography / Sandie Lindsay.—2nd ed.
        p.   cm.—(Analytical Chemistry by Open Learning)
    Includes bibliographical references (p.   ) and index.
    ISBN 0 471 93180 2 (cloth)   :   ISBN 0 471 93115 2
(paper)
    1. High performance liquid chromatography—Programmed instruction.
2. Chemistry, Analytic—Programmed instruction.   I. ACOL (Project)
II. Title.   III. Series: Analytical Chemistry by Open Learning
(Series)
QD79.C454L54   1992
543'.0894—dc20   91–428293
                                                          CIP

British Library Cataloguing in Publication Data:

A catalogue record for this book is available from the British Library

ISBN 0 471 93180 2 (cloth)
ISBN 0 471 93115 2 (paper)

Typeset in 11/13pt Times by Text Processing Dept, John Wiley & Sons Ltd, Chichester
Printed and bound in Great Britain by Courier International Ltd, East Kilbride, Scotland

The ACOL Series

This series of easy to read, user-friendly texts has been written by some of the foremost lecturers in Analytical Chemistry in the United Kingdom. The texts are designed for training, continuing education and updating of all technical staff concerned with Analytical Chemistry.

These texts are for those interested in Analytical Chemistry and instrumental techniques who wish to study in a more flexible way than traditional institute attendance or to augment such attendance.

Analytical Chemistry by Open Learning, ACOL, provides training resources—books, courses and computer programs. Courses based on ACOL, incorporating practical workshops and tutor assessment, are provided by Thames Polytechnic in conjunction with the Royal Society of Chemistry. BTEC qualifications are awarded for successful completion of courses and lead to a diploma that would be an appropriate preparation for applicants to the Royal Society of Chemistry for Licentiateship status, LRSC.

Thames Polytechnic will continue to support these 'Open Learning Texts', to continually refresh and update the material and to extend their coverage. For further information on ACOL, the RSC courses and the BTEC diploma, please contact:

The ACOL Office,
Thames Polytechnic,
Avery Hill Road,
Eltham,
London
SE9 2HB,
U.K.

How to Use an Open Learning Text

Open learning texts are designed as a convenient and flexible way of studying for people who, for a variety of reasons, cannot use conventional education courses. You will learn from this text the principles of one subject in Analytical Chemistry, but only by putting this knowledge into practice, under professional supervision, will you gain a full understanding of the analytical techniques described.

To achieve the full benefit from an open learning text you need to plan your place and time of study.

- Find the most suitable place to study where you can work without disturbance.

- If you have a tutor supervizing your study discuss with him, or her, the date by which you should have completed this text.

- Some people study perfectly well in irregular bursts, however most students find that setting aside a certain number of hours each day is the most satisfactory method. It is for you to decide which pattern of study suits you best.

- If you decide to study for several hours at once, take short breaks of five or ten minutes every half hour or so. You will find that this method maintains a higher overall level of concentration.

Before you begin a detailed reading of the text, familiarize yourself with the general layout of the material. Have a look at the course contents list at the front of the book and flip through the pages to get a general impression of the way the subject is dealt with. You will find that there is space on the pages to make comments alongside the text as you study—your own notes for highlighting points that you feel are particularly important. Indicate in the margin the points you would like to discuss further with a tutor or fellow student. When you come to revise, these personal study notes will be very useful.

Π When you find a paragraph in the text marked with a symbol such as is shown here, this is where you get involved. At this point you are directed to do things: draw graphs, answer questions, perform calculations, etc. Do make an attempt at these activities. If necessary cover the succeeding response with a piece of paper until you are ready to read on. This is an opportunity for you to learn by participating in the subject and although the text continues by discussing your response, there is no better way to learn than by working things out for yourself.

We have introduced self assessment questions (SAQs) at appropriate places in the text. These SAQs provide for you a way of finding out if you understand what you have just been studying. There is space on the page for your answer and for any comments you want to add after reading the author's response. You will find the author's response to each SAQ at the end of each Chapter. Compare what you have written with the response provided and read the discussion and advice.

At intervals in the text you will find a Summary and List of Objectives. The Summary will emphasize the important points covered by the material you have just read and the Objectives will give you a checklist of tasks you should then be able to achieve.

You can revise the Unit, perhaps for a formal examination, by re-reading the Summary and the Objectives, and by working through some of the SAQs. This should quickly alert you to areas of the text that need further study.

Contents

Study Guide

High performance liquid chromatography (HPLC) is the most powerful of all the chromatographic techniques. It can often easily achieve separations and analyses that would be difficult or impossible using other forms of chromatography. On the other hand, there are very many things that can go wrong with such separations; there are probably more pitfalls in HPLC than in any other form of chromatography. To avoid these, you have to have the sort of experience that is difficult to obtain by reading textbooks. Only by doing a great deal of experimental work (and making many mistakes) can you hope to achieve the necessary practical skills.

This is not to say that the theoretical side of the subject is unimportant. In chromatography, theory has always led experimental work. The great advances that have been made in liquid chromatography in the last 10–15 years have been achieved through a better theoretical understanding of the technique. You will not be able to use HPLC to full advantage unless you have a proper understanding of how it works. Liquid chromatography is a very wide ranging subject and to understand it you will have to have some knowledge of many different areas of physical and analytical chemistry. I will assume you have studied chemistry up to the standard of the BTEC higher certificate and that you have a knowledge of physics and mathematics up to about 'A' level standard. I will also assume that you are familiar with *ACOL: Chromatographic Separations*. You will find it helpful to have had some experience with the use of analytical instruments such as spectrophotometers and chart recorders.

As well as dealing with the basic principles of the method, I have attempted to give a brief coverage of one or two more specialized topics, for instance in Sections 6.5, 10.2 and 10.6. If you are new to chromatography and find these too difficult, you can treat them as optional if you wish.

Because HPLC covers such a wide area, you are bound to find that there are some topics that you would like to study in more detail than is given in this text. Suitable books that you could use as a starting point are listed in the bibliography. Because HPLC is a technique that is still developing, textbooks often contain some material that is obsolete. To get up to date information, especially on columns and instrumentation, you have to use the chromatographic literature, or catalogues and applications literature from equipment manufacturers. A selection of applications literature available from manufacturers is given at the end of Chapter 8.

Supporting Practical Work

1. GENERAL CONSIDERATIONS

The experiments below use reverse phase chromatography with bonded silica columns and uv absorbance detection. If more extensive experimental facilities are available, some additional experiments are suggested. These are concerned with the preparation and evaluation of columns, and with the use of other detectors and modes of hplc. It should be possible to complete each experiment within a three hour practical period.

2. AIMS

(a) To provide practical experience in the use of basic hplc equipment.

(b) To demonstrate the various parameters that control hplc separations.

(c) To show the use of the technique for separations and quantitative analysis.

(d) To illustrate some of the important principles from the theory part of the Unit.

3. SUGGESTED EXPERIMENTS

(a) The effect of mobile phase flow rate and dead volume on column performance.

(b) The effect of mobile phase composition on retention and selectivity in a reverse phase separation.

(*c*) Determination of 4-hydroxy-3-methoxy benzaldehyde (vanillin) in vanilla essence.

(*d*) Determination of aspirin and caffeine in an analgesic tablet.

4. ADDITIONAL EXPERIMENTS

(*a*) Preparation and evaluation of an hplc column.

(*b*) Analysis of sugars in fruit juice.

(*c*) The use of extraction techniques in the separation of carotene pigments from fruit.

Bibliography

Textbooks on high performance liquid chromatography.

1. J.J. Kirkland and L.R. Snyder, *Introduction to Modern Liquid Chromatography*, 2nd edn., Wiley, 1979.

2. R.J. Hamilton and P.A. Sewell, *Introduction to High Performance Liquid Chromatography*, Chapman and Hall, 1982.

3. C.F. Simpson, Ed. *Techniques in Liquid Chromatography*, Wiley, 1984.

4. J.H. Knox, Ed. *High Performance Liquid Chromatography*, Edinburgh University Press, 1982.

5. Veronika R. Meyer, *Practical High Performance Liquid Chromatography*, Wiley, 1988.

Reference 1 is a very comprehensive treatment. References 2, 4, 5 and parts of 3 are simpler treatments which are more suited to this text.

Useful material on HPLC can also be obtained from manufacturers catalogues and applications notes (see Chapter 8, after the Summary) and from a number of free journals including:

LC–GC International, Chester Business Park, Chester, CH4 9QH. Chromatography and Analysis, 27 Norwich Road, Halesworth, Suffolk, IP19 8BX.

Laboratory Equipment Digest, 30 Calderwood Street, London SE18 6QH.

Additional reading for specialised topics is given at the end of chapters.

Acknowledgements

I am grateful for assistance from Alan Curtis and Cecil Lobo (Bush Boake Allen Ltd), John Mills (Varian Associates), Tony Green (Eurocolour Ltd), Ken Evans (ICI Colours and Fine Chemicals), Dave Cook (Dyson Instruments) and, especially, from Tom Donovan (Biotage).

Figure 1.2 is redrawn from *Jones Chromatography 1989 Catalogue* with the permission of Jones Chromatography Ltd.

Figures 2.3d and 6.5d and e, are redrawn from *Chromatographia* 1982, 15, 693 with the permission of Friedr. Vieweg and Sohn Verlag.

Figure 3.1a is redrawn from *Anachem Chromatography Accessories*, 1990, with the permission of Kontes Glass Co.

Figure 3.5 is redrawn from *Rheodyne Tips on L C Injection*, 12/86, with the permission of Rheodyne Inc.

Figure 4.1c is redrawn from *Upchurch Scientific Catalogue*, 1989, with the permission of Upchurch Scientific.

Figure 5.4c has been redrawn from *Varian Instruments at Work*, 1987, uv-41, with the permission of Varian International.

Figures 5.5b, 7.5d and 7.6j are redrawn from *Perkin Elmer Publications*. Permission has been requested.

Figure 6.3b is redrawn from *Journal of Chromatographic Science* 1978, 16, 227. Permission has been granted by Preston Publications.

Figure 6.5c is supplied by and reproduced with the permission of ICI Colours and Fine Chemicals Research.

Figures 6.5g–k are redrawn from *LC-GC International*, 1990, 3, 54. Permission has been requested.

Figures 7.4a and 7.5e are redrawn from *FSA Catalogue*, 1990. Permission has been requested.

Figure 7.5b is redrawn from *Waters Source Book 1986*, with the permission of the Millipore Corporation.

Figure 7.5c is redrawn from *Journal of Chromatographic Science*, 1980, 18, 519. Permission has been requested.

Figure 7.6g is supplied and reproduced with the permission of Dyson Instruments Ltd.

Figures 7.6h, 7.6i, 7.7a, 7.7b, 7.7g, 8.2b–8.2f, 8.3c, 8.3d–8.3h are redrawn from *Waters LC School* with the permission of the Millipore Corporation.

Figure 8.1a is redrawn from *Alltech Associates Catalogue*, 1989, with the permission of Alltech Associates Inc.

Figures 8.4a and b, are supplied by and reproduced with the permission of Bush Boake Allen Ltd.

Figure 8.4i is redrawn from *Philips Analytical Advances LC2*. 1989. Permission has been requested.

Figure 10.1b is redrawn from *Supelco HPLC Bulletin 819*. Permission applied for.

Figure 10.3 has been redrawn from *May and Baker Technical Information*. Permission has been requested.

Figure 10.4b has been redrawn from *Waters Preparative Chromatography Notes*, 1, with the permission of Waters Internal Publication.

1. Introduction

1.1. HISTORY AND BASIC PRINCIPLES

High performance liquid chromatography (HPLC) is a technique that has arisen from the application to liquid chromatography (LC) of theories and instrumentation that were originally developed for gas chromatography (GC).

Classical liquid chromatography has been around for quite a long time, and you will probably have used it in one form or another. In the original method an adsorbent, for instance alumina or silica, is packed into a column and is eluted with a suitable liquid. A mixture to be separated is introduced at the top of the column and is washed through the column by the eluting liquid. If a component of the mixture (a solute) is adsorbed weakly on to the surface of the solid stationary phase it will travel down the column faster than another solute that is more strongly adsorbed. Thus separation of the solutes is possible if there are differences in their adsorption by the solid. This method is called adsorption chromatography or liquid solid chromatography (LSC).

In LC there are other sorption mechanisms that can cause separation, depending on whether we choose to use a liquid or a solid as the stationary phase, or what kind of solid we use. Liquid–liquid chromatography (LLC) uses a liquid stationary phase coated on to a finely divided inert solid support. Separation here is due to differences in the partition coefficients of solutes between the stationary liquid and the liquid mobile phase. In normal phase LLC the stationary phase is relatively polar and the mobile phase relatively non-polar, whilst reverse phase LLC uses a non-polar stationary liquid and a polar mobile phase.

In ion exchange chromatography the stationary phase is an ion exchange resin, and separations are governed by the strength of the interactions between solute ions and the exchange sites on the resin. Finally, in exclusion chromatography the stationary phase is a wide pore gel that can separate molecules on the basis of their size and shape, the largest molecules travelling most rapidly through the system.

Experimentally, classical LC was done by packing the stationary phase into a glass column, maybe 5 cm in diameter and 1 m in length, and eluting with a suitable solvent, or range of solvents. The column could often be used only once, having to be repacked for each sample that was examined. In LLC the eluting solvent had to be saturated with the stationary liquid phase, in order to avoid stripping the stationary liquid from the column. Many of the stationary phases used were not very efficient, so that for tricky separations long columns had to be used, the separations took a long time and used large amounts of solvent. The separated solutes were isolated by dividing the output of the column up into a series of arbitrary fractions, which were then evaporated down so that any solutes present could be identified by other physical or chemical methods (e.g. melting point, elemental analysis, spectrometry).

The development of the open-column methods, i.e. paper chromatography (in the 1940s) and thin layer chromatography (in the 1950s), greatly improved the speed and resolution of LC, but there were still serious limitations compared to modern LC methods, in that analysis times were long, resolution was poor and quantitative analysis, preparative separations and automation were difficult.

It was known from gas chromatographic theory that efficiency could be improved if the particle size of the stationary phase materials used in LC could be reduced. High performance liquid chromatography arose gradually in the late 1960s as these high efficiency materials were produced, and as improvements in instrumentation allowed the full potential of these materials to be realized. As HPLC has developed, the particle size of the stationary phase materials used in LC has become progressively smaller. The stationary phases used today are called microparticulate column packings and are commonly uniform, porous silica particles, with spherical or irregular shape, and nominal diameters of 10, 5 or 3 μm. The different separation mechanisms mentioned earlier can be realized by the bonding of different chemical groups to the surface of the silica particle, to produce what are called bonded phases. Chromatography suppliers list a variety of these, but about 75% of the work in HPLC at the moment is done using a bonded phase in which C-18 alkyl groups are attached to the surface of the silica particles. These types are called ODS (octadecylsilane) bonded phases. With bonded phases, the nature of the sorption mechanism is sometimes not clear, and there is much theoretical and experimental work going on at the moment attempting to clarify such mechanisms.

When packed into a column, the small size of these particles leads to a considerable resistance to solvent flow, so that the mobile phase has to be pumped through the column under high pressure. Typically, the column is 10–25 cm long and 4.6 mm in internal diameter. Although these columns are expensive, they are re-usable, so that the cost can be spread over a large number of samples. The column and all the associated plumbing must be able to withstand the pressures that are used, and must also be chemically resistant to the mobile phase solvents. Columns are usually made of stainless steel, although glass or plastics are used by some manufacturers.

In analytical HPLC the mobile phase is normally pumped through the column at a flow rate of 1–5 cm^3 min^{-1}. If the composition of the mobile phase is constant, the method is called 'isocratic' elution. Alternatively, the composition of the mobile phase can be made to change in a predetermined way during the separation, which is a technique called 'gradient' elution. Gradient elution is used in situations similar to those requiring temperature programming in GC, and is necessary when the range of retention times of solutes on the column is so large that they cannot be eluted in a reasonable time using a single solvent or solvent mixture. In adsorption chromatography, for instance, non-polar solutes are adsorbed relatively weakly and should be eluted with a non-polar solvent, whereas polar solutes are adsorbed more strongly and require a more polar solvent. If the sample contains a wide range of polarities, the separation could be done by changing the polarity of the solvent mixture during the separation. In other cases it may be necessary to use gradient elution where other properties of the solvent (e.g. pH or ionic strength) are changed.

After passing through the column, the separated solutes are sensed by an in-line detector. The output of the detector is an electrical signal, the variation of which is displayed on a potentiometric recorder, a computing integrator or a VDU screen. Most of the popular detectors in HPLC are selective devices, which means that they may not respond to all of the solutes that are present in the mixture. At present there is no universal detector for

HPLC that can compare with the sensitivity and general convenience of the flame ionization detector in gas chromatography. Some solutes are not easy to detect in HPLC, and have to be converted into a detectable form after they emerge from the column. This approach is called 'post-column derivatization'.

As in other forms of chromatography, the time taken for the solute to pass through the chromatographic system (the retention time) is a characteristic of the solute for a particular set of conditions. However, to use retention data on their own for the identification of unknown solutes would be rather like trying to identify an unknown organic compound by simply measuring its melting point or boiling point. Many different solutes will have identical retention times for a particular set of conditions. Chromatography is an excellent method for the separation of mixtures but it does not provide the detail necessary for the clear identification of the separated compounds. Such detail is provided by spectrometric techniques, so it is not surprising that a lot of effort has been expended to try and combine them with HPLC. For example, some detectors can record and store the UV spectra of solutes as they emerge from the column. A much more powerful (and very expensive) method is the direct combination of liquid chromatography and mass spectrometry. In both cases, modern data processing methods allow you to match the spectra obtained with the spectra of standard substances, obtained from libraries of recorded spectra.

How do we decide to separate a mixture by GC or HPLC? In GC, mixtures are examined in the vapour phase, so that we have to be able to form a stable vapour from our mixture, or convert the substances in it to derivatives that are thermally stable. Only about 20% of chemical compounds are suitable for GC without some form of sample modification; the remainder are thermally unstable or involatile. In addition, substances with highly polar or ionizable functional groups often show poor chromatographic behaviour by GC, being very prone to tailing. For HPLC, the only restriction is that we have to be able to dissolve our sample in a solvent. Thus HPLC is the better method for macromolecules, inorganic or other ionic species, labile natural products, pharmaceutical compounds and biochemicals.

In GC there is only one phase, the stationary liquid or solid phase, that is available for interaction with the sample molecules. Because the mobile phase is a gas, all sample vapours are miscible with it in all proportions. In HPLC both the stationary phase and the mobile phase can interact selectively with the sample. Interactions such as complex.... hydrogen bonding that are absent in the GC mobile phase may occur in the HPLC mobile phase. The variety of these selective interactions can also be increased by suitable chemical modification of the silica surface. Because of all this, HPLC is a more versatile method than GC, and can often achieve much more difficult separations.

There will often be areas where either method could be used, and in such cases GC is usually chosen. One reason for this is that HPLC tends to be a more expensive method than GC, both in capital outlay for the instrument and in day-to-day running costs. The GC separation is also often faster and more sensitive.

1.2. BASIC INSTRUMENTATION

Although HPLC equipment is discussed fully in Chapters 3–5 the basic components are dealt with briefly in this section. You will have seen from reading Section 1.1 that an HPLC instrument requires a high pressure pump and a supply of mobile phase, a column containing a high efficiency stationary phase, an injection unit for introducing samples on to the column, an in-line detector and some method of displaying the detector signal. Figure 1.2a is a block diagram showing the way in which these different components are arranged to form a high performance liquid chromatograph.

The mobile phase in HPLC may be water, organic solvents or buffers either on their own or mixed with one another. Any part of the system that is in contact with the mobile phase must be made of materials that are not attacked by any of the solvents that are to be used. The wetted parts are usually made of stainless steel, PTFE or other inert plastics, although other materials are sometimes used in the pumps, e.g. sapphire, ruby or ceramics. Everything on the high pressure side, i.e. from the pump outlet to the end of the column, must be strong enough to withstand the pressures involved.

Another important design consideration is that, between the point at which the sample is introduced and the point at which it is detected, the dead volume in the equipment must be kept to a minimum. Dead volume means any empty space or unoccupied volume, the presence of which can lead to disastrous losses in efficiency (Figs. 2.5a and 9.3b show examples of this). Clearly there will be some dead volume in the column itself, which will be

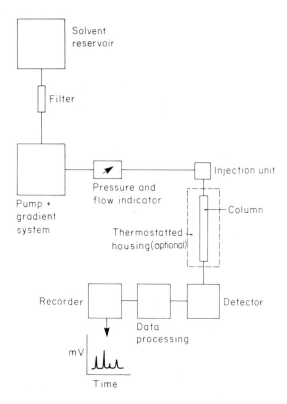

Fig. 1.2a. *Block diagram of a high performance liquid chromatograph*

the space that is not occupied by the stationary phase. The other sources of dead volume are the injection unit, the tubing and fittings at each end of the column, and the detector cell. We should therefore use small bore tubing in short lengths for making the injector–column and column–detector connections, and the injector and detector must be designed so that their internal volume is as small as possible. Dead volume before the introduction of the sample should also be minimized, as this reduces the time needed to change the mobile phase composition.

Figure 1.2b is a reverse phase separation of a test mixture using a bonded silica stationary phase, obtained on a modern instrument, and shows the speed at which separation of a complex mixture can be achieved when conditions are favourable.

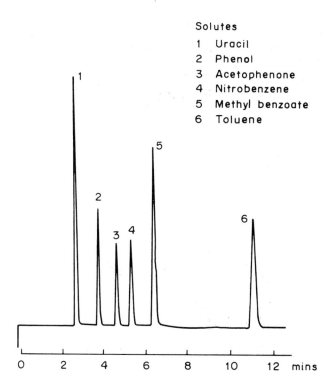

Fig. 1.2b. *Chromatogram of test mixture*

Column: *Apex 5 μm C-18, 25 cm × 4.6 mm*

Mobile phase: *CH₃OH/H₂O 65:35*

Flow rate: *0.5 cm³ min⁻¹*

Detector: *UV absorption, 254 nm*

SAQ 1a

Choose one of the options (*i*)–(*iv*) for each of the following:

1. A normal phase separation in HPLC implies that:

   (*i*) A bonded silica is used as the stationary phase;

   (*ii*) The mobile phase is more polar than the stationary phase;

   (*iii*) The mobile phase is less polar than the stationary phase;

   (*iv*) A non-polar solvent is used for the mobile phase.

2. In liquid chromatography the use of a single eluent for the whole separation is called:

   (*i*) Normal elution;

   (*ii*) Reverse elution;

   (*iii*) Isocratic elution;

   (*iv*) Frontal elution.

3. Gradient elution is used for:

   (*i*) Mixtures of polar and non-polar compounds;

   (*ii*) Mixtures of compounds with large differences in molecular mass;

   (*iii*) Mixtures of compounds with different volatility;

$\longrightarrow$

SAQ 1a
(cont.)

> (*iv*) Closely related compounds.
>
> 4. Derivative preparation in HPLC is used for:
>
>   (*i*) Improving sensitivity;
>
>   (*ii*) Compounds that are highly polar;
>
>   (*iii*) Compounds that are involatile;
>
>   (*iv*) Compounds that are thermally unstable.

SAQ 1b

Complete the following definition of liquid chromatography by filling in the blanks. For each space, choose a word from the groups given below.

Liquid chromatography is a method for the .... of mixtures in which the sample is introduced into a system of two .... Differences in .... shown by the solutes cause them to travel at different speeds in the ....

(*i*) ANALYSIS   (*iii*) ADSORPTION
     SEPARATION   DISTRIBUTION
     DETERMINATION   PARTITION

(*ii*) SUBSTANCES   (*iv*) LIQUID
      CHEMICALS   MOBILE PHASE
      PHASES   SYSTEM

SAQ 1c

For which of the following would HPLC be a suitable method of analysis?

(i) Determination of the composition of cigarette lighter fuel;

(ii) Analysis of ascorbic acid (vitamin C) in a vitamin C tablet;

(iii) Determination of the amount of caffeine in a soft drink;

(iv) Separation of a mixture of naturally occurring sugars;

(v) Separation of a mixture of amines.

1.3. SUMMARY

A brief description is given of the way in which modern liquid chromatography has been developed from classical techniques. The important components of a high performance liquid chromatograph are discussed briefly.

Learning Objectives

Now that you have completed Chapter 1 you should be able to:

• List the mechanisms by which separations are achieved in HPLC;

• Appreciate that a very wide range of samples can be separated using the technique;

• Understand the arrangement of the main components of a high performance liquid chromatograph;

• Identify typical materials used in HPLC equipment;

• Recognize the need for minimal dead volume in the equipment and identify the sources of dead volume.

SAQs AND RESPONSES

SAQ 1a	Choose one of the options (i)–(iv) for each of the following:
	1. A normal phase separation in HPLC implies that:
	(i) A bonded silica is used as the stationary phase;
	(ii) The mobile phase is more polar than the stationary phase;
	—⟩

SAQ 1a
(cont.)

(*iii*) The mobile phase is less polar than the stationary phase;

(*iv*) A non-polar solvent is used for the mobile phase.

2. In liquid chromatography the use of a single eluent for the whole separation is called:

(*i*) Normal elution;

(*ii*) Reverse elution;

(*iii*) Isocratic elution;

(*iv*) Frontal elution.

3. Gradient elution is used for:

(*i*) Mixtures of polar and non-polar compounds;

(*ii*) Mixtures of compounds with large differences in molecular mass;

(*iii*) Mixtures of compounds with different volatility;

(*iv*) Closely related compounds.

4. Derivative preparation in HPLC is used for:

(*i*) Improving sensitivity;

(*ii*) Compounds that are highly polar;

(*iii*) Compounds that are involatile;

(*iv*) Compounds that are thermally unstable.

Response

1. (*iii*); 2. (*iii*); 3. (*i*); 4. (*i*).

****

SAQ 1b

Complete the following definition of liquid chromatography by filling in the blanks. For each space, choose a word from the groups given below.

Liquid chromatography is a method for the .... of mixtures in which the sample is introduced into a system of two .... Differences in .... shown by the solutes cause them to travel at different speeds in the ....

(*i*) ANALYSIS   (*iii*) ADSORPTION
     SEPARATION   DISTRIBUTION
     DETERMINATION   PARTITION

(*ii*) SUBSTANCES   (*iv*) LIQUID
      CHEMICALS   MOBILE PHASE
      PHASES   SYSTEM

Response

(*i*) ANALYSIS or DETERMINATION are not too bad, but the real power of chromatography is as a separation method, so SEPARATION is the one I would choose. (ii) PHASES emphasizes the way the method works, the other two are too vague. (iii) DISTRIBUTION is best (the other two are different kinds of sorption mechanisms and are too specific). (iv) LIQUID is not correct, as both the mobile and the stationary phase may be liquids. MOBILE PHASE is not right either, as the solutes all travel at the same speed when they are in the mobile phase. So you are left with SYSTEM, meaning 'combination of stationary and mobile phases'.

****

SAQ 1c	For which of the following would HPLC be a suitable method of analysis? (i) Determination of the composition of cigarette lighter fuel; (ii) Analysis of ascorbic acid (vitamin C) in a vitamin C tablet; (iii) Determination of the amount of caffeine in a soft drink; (iv) Separation of a mixture of naturally occurring sugars; (v) Separation of a mixture of amines.

Responses

(*i*) This will be a mixture of light hydrocarbons and would be a clear case for GC, which would give an easier, quicker and cheaper separation than HPLC;

(*ii*) The tablet will contain ascorbic acid, together with insoluble fillers and binders. The only separation needed would be to take up the tablets in water and filter from insoluble material. The determination could be done by HPLC, but there are titrimetric or electrochemical methods that would be easier;

(*iii*) The caffeine may have to be separated from flavours, colouring materials and other additives present in the drink. For this one, HPLC would be a suitable technique;

(*iv*) and (*v*) Both of these could be separated by GC or by HLPC, but HPLC would probably be the better technique in both examples. The sugars could not be volatilized without decomposition, so by GC they would have to be examined as volatile TMS (trimethylsilyl) derivatives. The amines are polar substances that would show pronounced tailing by GC and would have to be derivatized as well. Both could be examined by HPLC without pretreatment.

****

2. Retention and Peak Dispersion

This section deals briefly with some of the ways that are used to describe the quality of a chromatographic separation, the factors that affect the quality, and the methods that can be used to improve it. We need to be able to describe where on the chromatogram our solute peaks are eluted, whether or not they are separated from one another, and if so how efficiently. We need to be able to express all these quantities as numbers.

2.1. RETENTION MEASUREMENTS

Fig. 2.1a shows the measurements that are taken from a chromatogram in order to quantify results. In the figure the two large peaks are from solutes that are retained on the column, and the small peak represents an unretained solute that travels through the column at the same speed as the mobile phase. The quantities t_{R1}, t_{R2} and t_0 are the retention times of these three solutes and can be measured as times (as in Fig. 2.1a), volumes of solvent or distances on a recorder chart. Although a peak in the chromatogram can be identified by its retention time, because this varies with column length and mobile phase flow rate, it is better to locate or identify peaks using the capacity factor (k'), which is given by

$$k' = \frac{t_R - t_0}{t_0} \qquad (2.1a)$$

In HPLC separations, we try to keep the capacity factors of our solutes between 1 and 10, since if k' values are too low it is likely that the solutes may not be adequately resolved, and for high k' values the analysis time may be too long.

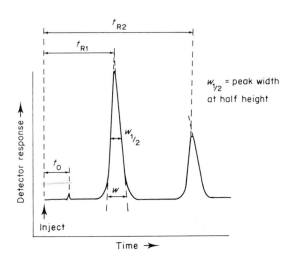

Fig. 2.1a.   *Retention and dispersion measurements*

The capacity factor tells us where the peaks elute relative to an unretained solute. The separation of the two peaks relative to each other is described by the selectivity or separation factor (α), which is defined for two peaks as the ratio of the capacity factors. By convention this equation is written so that $\alpha \geq 1$.

$$\alpha = \frac{k_2'}{k_1'} = \frac{t_{R2} - t_0}{t_{R1} - t_0} \qquad (2.1b)$$

The degree of separation of one component from another is described by the resolution (R_S), measured as the difference in retention time of the two solutes divided by their average peak width.

$$R_S = \frac{t_{R2} - t_{R1}}{0.5(w_1 + w_2)} \qquad (2.1c)$$

The values of t and w here can again be measured in volume, time or distance units as long as we use the same units for each. When two peaks are just resolved to the baseline this corresponds to $R_s = 1.5$. Figure 2.1b shows how the measurement is made for a pair of partly resolved peaks.

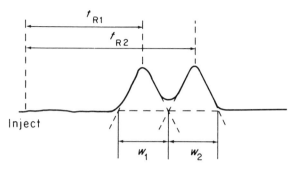

Fig. 2.1b. *Measurement of resolution*

2.2. COLUMN PERFORMANCE

One of the problems with any form of chromatography is that as a band of solute moves through the system it becomes dispersed; the longer the solute spends in the system the more dispersed it becomes.

However, the more efficient the chromatographic column, the less of this band spreading will occur. In Fig. 2.1a. the more efficient the column, the smaller will be w_1 and w_2 at a given value of t_{R1} or t_{R2}. To measure efficiency we use quantities called the plate number (N) or plate height (H), which are defined as follows

$$N = 16 \left(\frac{t_R}{w} \right)^2 \tag{2.2a}$$

or

$$N = 5.54 \left(\frac{t_R}{w_{1/2}} \right)^2 \tag{2.2b}$$

$$H = \frac{L}{N} \tag{2.2c}$$

where L = column length.

Of the two measurements, the plate height is generally preferred, as it measures the efficiency for unit column length, and is therefore useful for comparative purposes. The plate number can be increased by simply increasing the length of the column.

The plate number is dimensionless, but H has units of length. Generally the efficiency of HPLC columns increase as the particle size of the packing decreases. Commercial reverse phase columns using bonded silica have about 50 000 plates m^{-1} when packed with 5 μm particles and about 25 000 plates m^{-1} for 10 μm particles. The plate number that you need depends very much on the sort of work you are doing; much routine work in HPLC is done at efficiencies considerably lower than these.

Π   What would be the typical plate height of a 12.5 cm reverse phase HPLC column with a 5 μm packing?

Roughly, $N = \dfrac{50000}{8} = 6250$

$H = \dfrac{125}{6250} = 0.02$ mm

2.3.   THE RESOLUTION EQUATION

This equation shows the dependence of resolution on selectivity, capacity factor and plate number for two peaks and is the key equation for the optimization of resolution in chromatography. Resolution may be defined as

$$R_S = 0.25 \left(\frac{\alpha - 1}{\alpha} \right) \left(\frac{\bar{k}}{1 + \bar{k}} \right) N^{1/2} \qquad (2.3a)$$

where $\bar{k}$ is the average capacity factor for the two peaks and N is the average plate number.

The equation shows that, for a desired degree of resolution, three conditions have to be met:

(a) The peaks must be separated from each other ($\alpha > 1$);

(b) The peaks must be retained on the column ($\bar{k} > 0$);

(c) The column must develop some minimum number of plates.

The influence of each of the three factors, α, $\bar{k}$ and N on the resolution can be discussed independently of the other two. If we assume constant selectivity and efficiency terms, resolution is proportional to the factor $\bar{k}/(1+\bar{k})$, the variation of which is shown in Fig. 2.3a(i). The limiting value of the term $\bar{k}/(1+\bar{k})$ is 1. If we increase $\bar{k}$, the resolution between peaks increases significantly at first, but the effect diminishes at higher values of $\bar{k}$. When $\bar{k} = 5$ we have reached 83% and, for $\bar{k} = 9$, 90% of the limiting value. The optimum values for $\bar{k}$ are in the range 1–10; very little can be gained by increasing $\bar{k}$ further. High values of $\bar{k}$ also mean long analysis times. The capacity factor is controlled largely by adjustment of the mobile phase composition.

Since $\alpha > 1$ the limiting value of the term $(\alpha - 1)/\alpha$ is 1 also. This term increases rather more regularly with increase in α than does the first term

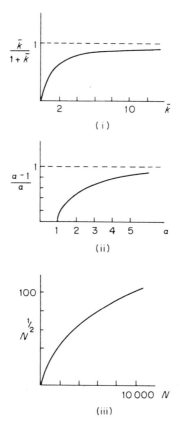

Fig. 2.3a. *Effect of $\bar{k}$, α and N on resolution*

with increase in $\bar{k}$. When $\alpha = 2$ (a large selectivity in practical terms) we have reached only 50% of the limiting value; see Fig. 2.3a(ii). The value of R_S is most sensitive to α when its value is close to one, but also at higher values resolution will benefit from an increase in the value of α. Selectivity can be altered by changing either the nature of the stationary phase, or the nature or composition of the mobile phase.

In the variation of resolution with plate number, the square root function initially rises steeply from zero near $N = 0$, see Fig. 2.3a(iii), but then increases less rapidly at higher values of N. To double the resolution, N must be increased by a factor of 4, which is difficult to achieve in practice. The plate number can be changed by altering the length of the column, the particle size of the packing or the flow rate of the mobile phase. If we simply increase N by increasing the column length, this leads to longer analysis times and a greater pressure drop across the column. Unless the column is obviously inferior, N is probably the least rewarding factor for increasing resolution.

Fig. 2.3b shows what happens to two partly resolved peaks when $\bar{k}$, N or

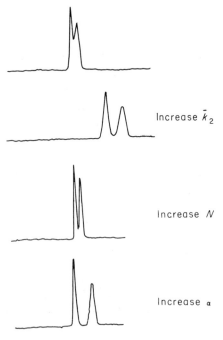

Increase $\bar{k}_2$

Increase N

Increase α

Fig. 2.3b.   *Effect on chromatogram of change of $\bar{k}$, N or α*

α is changed. If we increase N, other things being equal, the solutes appear at the same places in the chromatogram but the resolution is better because the width of each peak has decreased. Increasing $\bar{k}$ improves resolution by causing the solutes to spend more time in the stationary phase (but this increases analysis time). Increasing the selectivity moves one or both peaks relative to the other.

In practice, if we change α or $\bar{k}$ then one or both of the other two factors may change as well, so there is often a certain amount of trial and error involved in developing an HPLC method, as the examples in Chapter 8 will show.

SAQ 2.3a

The chromatogram in Fig. 2.3c has some partly resolved peaks. Assume that the first peak is an unretained solute and take t_0 as the position where the first peak starts to elute.

(*i*) Draw a scale on the abscissa marking k' values from 0 to 5.

(*ii*) Determine k' for each peak.

(*iii*) Measure peak widths and calculate the resolution between peaks 1 and 2, peaks 2 and 3 and peaks 4 and 5. For the peaks that are partly resolved, you will have to extrapolate the linear part of the side of each peak down to the baseline, as in Figure 2.1b.

(*iv*) Calculate the selectivity for peaks 2 and 3, 3 and 4, 4 and 5.

(*v*) For the last two peaks, calculate the plate number and the plate height of the column using Eqs. 2.2b and 2.2c (the column is 25 cm long and is not very efficient).

$\longrightarrow$

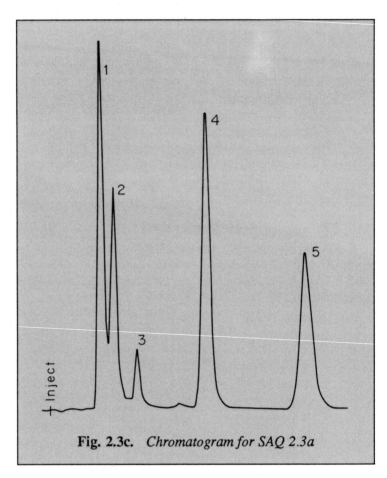

Fig. 2.3c. *Chromatogram for SAQ 2.3a*

In the chromatogram shown in Fig. 2.3d the retention times for peaks 4 and 6 are 354 and 373 s respectively. The column has 3500 plates and t_0 for the system is 123 s.

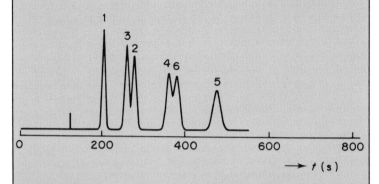

Fig. 2.3d. *Chromatogram for SAQ 2.3b*

(*i*) Calculate the resolution between peaks 4 and 6 using Eq. 2.3a;

(*ii*) Using the same mobile phase, what plate number is needed to produce unit resolution for these two peaks?

(*iii*) If we increase retention by using a mobile phase with more water, assuming that α and N remain the same as in (i): calculate k' for peaks 4 and 6 when the resolution between them = 1;

(*iv*) How long would it take to elute peak 6 with the conditions used in (iii).

SAQ 2.3b

SAQ 2.3c

A mixture of two substances is to be separated on a preparative scale using a column with $N = 1000$. A resolution of 1.2 is required. The mixture is first tested on an analytical column with $N = 7000$. Assuming that capacity factors are the same on both columns, what is the minimum resolution required on the analytical column to ensure satisfactory resolution on scaling up?

2.4. COLUMN DISPERSION MECHANISMS

There are three mechanisms that produce dispersion of a band of solute as it travels through the column:

(*a*) Eddy diffusion and flow dispersion is the term for the dispersion produced because of the existence of different flow paths by which solutes can progress through the column. These path differences arise because the stationary phase particles have different size and shape and because the packing of the column is imperfect, causing gaps or voids in the column bed. If all solute molecules travelled at the same speed, those in different flow paths would travel different distances in a given time. In practice the solute travels faster in the wider flow paths than in the narrower ones. Even in the same flow path the solute will travel faster when in midstream than when close to the stationary phase particles (see Fig. 2.5b below). To reduce dispersion due to multiple path effects we need to pack the column with small particles with as narrow a size distribution as possible. Unfortunately, as the particle size is reduced, the size distribution becomes more difficult to control.

Lateral diffusion, which means the movement of solute in a radial direction across the column (thus moving from one flow path to another), reduces the multiple path dispersion, as it tends to equalize the speed of solute species in the column. The longer a solute species spends in the column the more lateral diffusion will occur, so flow dispersion is slightly flow-dependent, being reduced by the use of low mobile phase flow rates.

(*b*) Longitudinal diffusion. Dispersion also arises because of diffusion of solute in the longitudinal (axial) direction in the column. This is an important source of dispersion in gas chromatography but less so in liquid chromatography because rates of diffusion are very much slower in liquids than in gases (roughly 10^3–10^4 times smaller). This effect will become more serious the longer the solute species spends in the column, so, unlike flow dispersion, it is reduced by using a rapid flow rate of mobile phase.

(*c*) Mass transfer effects. These effects arise because the rate of the distribution process (sorption and desorption) of the solute species between mobile and stationary phases may be slow compared to the rate at which the solute is moving in the mobile phase. When solute species interact with the stationary phase, they may spend some time in or on the stationary phase before rejoining the mobile phase, and in this time they may have been left behind by those solute species which did not interact. The stationary phase particle has an open porous structure and the internal pores will contain immobile or stagnant mobile phase through which the solute will have to diffuse before it can get at the stationary phase. Solute that diffuses a long way into the porous structure will be left behind by solute that bypasses the particle or diffuses only a short distance into it.

∏   Do you think that these mass transfer effects would become more serious at low or high mobile phase flow rates?

In the time it takes for mass transfer into or out of the stationary phase, we want a non-interacting solute species to have travelled as little as possible in the column, so we need a low flow rate. Mass transfer effects will also be reduced by using small particles or particles with a thin porous surface layer. A low-viscosity mobile phase will also reduce this effect. Eddy diffusion and mass transfer effects are shown in Fig. 2.4a.

You can see that these dispersion mechanisms are affected in different ways by the flow rate of mobile phase. To reduce dispersion due to longitudinal diffusion we need a high flow rate, whereas a low flow rate is needed to reduce dispersion due to the other two mechanisms. This suggests that there will be an optimum flow rate where the combination of the three effects produces minimum dispersion. This can be observed in practice if N or H (which measure dispersion) is plotted against the velocity (u) or the flow rate of the mobile phase. The result is the well known Van Deemter plot (Fig. 2.4b) in which the plate height of the column is expressed as the sum of the various band spreading mechanisms. A simplified equation of the plot is:

$$H = A + \frac{B}{u} + Cu \qquad (2.4a)$$

A is a flow dispersion term, and independent of u (in practice in LC, A increases slightly with increase in u); B is a longitudinal diffusion term; C is a mass transfer term.

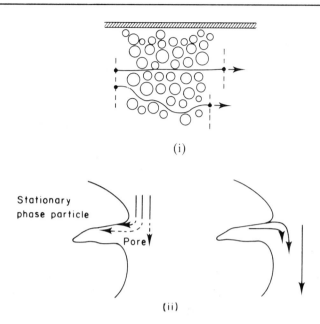

Fig. 2.4a.

(i) *Eddy diffusion*

Two molecules start at the same place in the column. If they both travel at the same speed, the molecule with the simpler flow path travels further in the column in a given time. Flow path differences may be greater in the wall regions of the column, where packing is irregular.

(ii) *Mass transfer*

Solute molecules which diffuse into the stationary phase particles and interact with them are left behind by those molecules that bypass the stationary phase

The slope of the resultant curve after the minimum depends on the particle size of the packing and on the column diameter. In analytical HPLC, where we are using small particle size packings in relatively narrow columns, the curve is often much flatter after the minimum than is suggested in the figure. This means that we can use relatively high flow rates (giving shorter analysis times) without sacrificing too much in loss of efficiency.

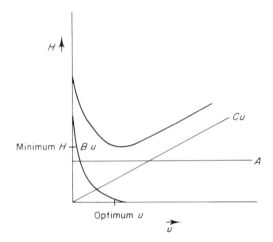

Fig. 2.4b. *Van Deemter plot. The top curve is the sum of the three band spreading mechanisms*

2.5. EXTRA-COLUMN DISPERSION

Dispersion can be produced outside the column by dead volumes in the injector, the detector or the plumbing. The combined effect of all of these is called extra-column dispersion. Figure 2.5a shows an example of this, in which different dead volumes are connected between the column and the detector, and Fig. 9.3b shows dispersion produced by dead volume at the top of the column. You can see from these that dead volume effects can cause a serious loss in performance.

We can examine the effect of dead volume by considering the peak width w in Fig. 2.1a as a volume and supposing it to be due to the dispersion produced by the column alone, i.e. what we would observe if the column was used in an ideal chromatograph that had no dead volume. Now imagine a real chromatograph without a column, in which the injector is connected to the column outlet tube. If we injected a solute we would observe a peak whose width w_A (measured as a volume) will be due only to extra-column effects. Most of this occurs as the solute flows through tubes, and leads to band spreading as shown in Fig. 2.5b.

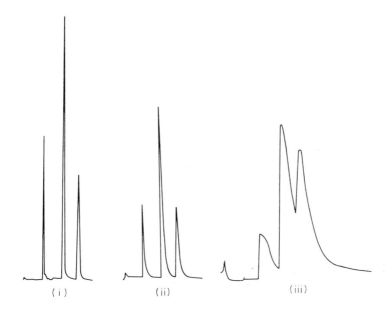

Fig. 2.5a. *Extra-column dispersion*

(i) Normal chromatogram
(ii) Chromatogram obtained with a 15 cm length of 0.8 mm i.d. tube inserted between column and detector inlet tube;
(iii) Chromatogram obtained with a 12.5 cm × 4.6 mm tube inserted as above.

Sample and conditions were as for the example in Section 8.4.3.

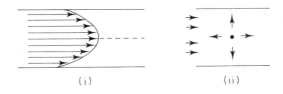

Fig. 2.5b. *Dispersion due to flow through tubes*

(i) The speed of the liquid varies over the cross-sectional area of the tube, and the solute molecules near the centre move faster than those near the walls
(ii) Dispersion can be caused by diffusional movement of the solute as the mobile phase flows through the tube

Suppose that we now use a column in the system and that the width of our solute peak is w_T. This will be larger than w because of extra-column dispersion. The relation between the three peak widths is:

$$w_T^2 = w^2 + w_A^2 \qquad (2.5a)$$

The value of w_A is likely to be 40–60 μl, unless the system has been specially designed for use with small bore columns. The effect on a chromatogram is more serious the earlier a peak is eluted, so we will examine the worst case, which is the effect on an unretained solute (one that travels at the same speed as the mobile phase).

Π   In a column packed with 5 μm silica, the solvent occupies about 70% of the volume of the column. For a 25 cm × 4.6 mm column packed with 5 μm silica, calculate:

(a) The retention volume (in μl) of an unretained solute;

(b) The width w (in μl) of the unretained solute peak. Use Eq. 2.2a and assume that the plate number of the column is 10 000.

(c) The width w_T that we would observe for this peak. Use Eq. 2.5a and assume that the extra-column dispersion is 50 μl.

Remember than 1 cm^3 = 1000 mm^3 = 1000 μl.

(*a*) The retention volume of an unretained solute is equal to the volume of solvent in the column.

Volume of empty column $\quad = \pi \times \left(\dfrac{4.6}{2}\right)^2 \times 250$

$$= 4155 \ \mu l$$

Volume of solvent in the column = 4155 × 0.7

$$= 2908.5 \ \mu l.$$

(b) $10\ 000 = 16 \times (2908.5/w)^2$

$\therefore w = 116.3\ \mu l.$

(c) $w_T^2 = 116.3^2 + 50^2$

$\therefore w_T = 126\ \mu l.$

The extra-column dispersion governs the dimensions of the column that we use. In the calculation above, the dispersion is increased by about 8% by the extra-column effects. If we want the dispersion to be increased by no more than this, then w should not be any smaller than the value calculated above. This in turn limits the retention volume, and thus the volume of the column itself. The minimum column volume we can use will depend on the amount of extra-column dispersion and on what we consider to be an acceptable increase in peak width that is produced by extra-column effects. In practice, this acceptable increase is taken as 10%, based on an unretained solute, and if we take 50 μl as a typical figure for extra-column dispersion then the minimum column diameter works out at about 4.5 mm for a column 25 cm long.

SAQ 2.5a

In Section 2.5 you worked out the effect of extra-column dispersion on the peak of an unretained solute, using a column with a plate number of 10 000. The extra-column dispersion will decrease the plate number that we actually observe for this column. The table below contains the retention volume, peak widths and plate number for an unretained solute on this column.

(a) Complete the table by calculating the corresponding values for columns with plate numbers of 5000 and 3000 respectively;

(b) What is the percentage reduction in the plate number of each column due to extra-column effects?

$\longrightarrow$

SAQ 2.5a
(cont.)

N (ideally)	10 000	5000	3000
V_R, μl	2908.5	2908.5	2908.5
w, μl	116		
w_T, μl	126		
N (actual)	8525		
% reduction in N			

2.6. SUMMARY

Retention, resolution and efficiency measurement are described and the resolution equation is discussed. Column dispersion mechanisms and the effects of extra-column dispersion are introduced.

Learning Objectives

Now that you have completed Chapter 2 you should be able to:

- Define capacity factor, selectivity and resolution, and measure these quantities from a chromatogram;

- Define plate number and plate height and measure these quantities from a chromatogram;

- Appreciate the effect of capacity factor, selectivity and plate number on resolution;

- Understand the operation of column dispersion mechanism;

- Describe the effect of dead volume on column performance.

References

1. C.F. Simpson, Ed. *Techniques in Liquid Chromatography*, Wiley, 1984, Chapters 1 and 2.

2. P.A. Sewell and B. Clarke, *ACOL Chromatographic Separations*, Wiley, 1987, Chapters 2 and 3.

3. J.H. Knox, Ed. *High Performance Liquid Chromatography*, Edinburgh University Press, 1982, Chapter 2.

4. P.J. Schoenmakers, *Optimisation of Chromatographic Selectivity*, Elsevier, 1986, Chapter 1.

SAQS AND RESPONSES

SAQ 2.3a   The chromatogram in Fig. 2.3c has some partly resolved peaks. Assume that the first peak is an unretained solute and take t_0 as the position where the first peak starts to elute.

(*i*) Draw a scale on the abscissa marking k' values from 0 to 5.

(*ii*) Determine k' for each peak.

(*iii*) Measure peak widths and calculate the resolution between peaks 1 and 2, peaks 2 and 3 and peaks 4 and 5. For the peaks that are partly resolved, you will have to extrapolate the linear part of the side of each peak down to the baseline, as in Figure 2.1b.

(*iv*) Calculate the selectivity for peaks 2 and 3, 3 and 4, 4 and 5.

(*v*) For the last two peaks, calculate the plate number and the plate height of the column using Eqs. 2.2b and 2.2c (the column is 25 cm long and is not very efficient).

$\longrightarrow$

SAQ 2.3a
(cont.)

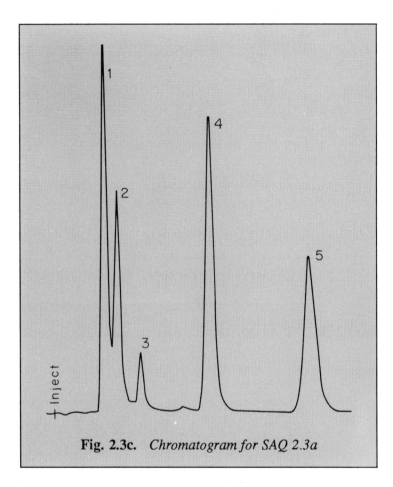

Fig. 2.3c. *Chromatogram for SAQ 2.3a*

Response

(*i*) *k'*		0	1	2	3	4	5
Distance from injection		12	24	36	48	60	72 mm

Peak no.	1	2	3	4	5
t_R, mm	14.5	18	24	43	70
(ii) k'	0.2	0.5	1.0	2.6	4.8
w_1 mm	3	3.5	3.5	4.5	7.0
$w_{1/2}$, mm				2.5	3.5
(iii) R_S		1.1	1.7	4.7	
(iv) α		2	2.6	1.8	
(v) N				1638	2216
H, mm				0.15	0.11

****

SAQ 2.3b

In the chromatogram shown in Fig. 2.3d the retention times for peaks 4 and 6 are 354 and 373 s respectively. The column has 3500 plates and t_0 for the system is 123 s.

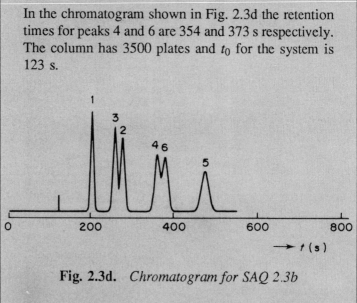

Fig. 2.3d. *Chromatogram for SAQ 2.3b*

(*i*) Calculate the resolution between peaks 4 and 6 using Eq. 2.3a;

$\longrightarrow$

SAQ 2.3b (cont.)	(ii) Using the same mobile phase, what plate number is needed to produce unit resolution for these two peaks?
	(iii) If we increase retention by using a mobile phase with more water, assuming that α and N remain the same as in (i): calculate k' for peaks 4 and 6 when the resolution between them = 1;
	(iv) How long would it take to elute peak 6 with the conditions used in (iii).

Response

We need to calculate capacity factors (Eq. 2.1a), selectivity (Eq. 2.1b) and then the resolution (Eq. 2.3a).

(i) $k_4' = \dfrac{354 - 123}{123} = 1.878$ $k_6' = \dfrac{373 - 123}{123} = 2.033$

$\alpha_{6,4} = \dfrac{2.033}{1.878} = 1.083$ $\bar{k} = \dfrac{1}{2}(k_4' + k_6') = 1.956$

$$R_S = 0.25 \times \frac{0.083}{1.083} \times \frac{1.956}{2.956} \times 3500^{1/2} = 0.01268 \times 3500^{1/2}$$

$$= 0.75$$

(ii) $1 = 0.01268N^{1/2}$ $\therefore$ $N = 6220$

(iii) For $R_S = 1$ $1.0 = 0.25 \times \dfrac{0.083}{1.083} \times \dfrac{\bar{k}}{1 + \bar{k}} \times 3500^{1/2}$

$\therefore \bar{k} = 7.49$

since $7.49 = \dfrac{1}{2}(k'_4 + k'_6)$ and $\dfrac{k_6'}{k_4'} = 1.083$

$k_4' = 7.19$ $k_6' = 7.79$

(*iv*) $7.79 = \dfrac{t_{R6} - 123}{123}$ $\therefore t_{R6} = 1081s$ (18 min)

****

SAQ 2.3c

A mixture of two substances is to be separated on a preparative scale using a column with $N = 1000$. A resolution of 1.2 is required. The mixture is first tested on an analytical column with $N = 7000$. Assuming that capacity factors are the same on both columns, what is the minimum resolution required on the analytical column to ensure satisfactory resolution on scaling up?

Response

For the two columns $1.20 = 0.25 \left(\dfrac{\alpha - 1}{\alpha} \right) \left(\dfrac{\bar{k}}{1 + \bar{k}} \right) \times 1000^{1/2}$

$$R_S = 0.25 \left(\dfrac{\alpha - 1}{\alpha} \right) \left(\dfrac{\bar{k}}{1 + \bar{k}} \right) \times 7000^{1/2}$$

Since k is the same on both columns, α must be the same also

$$\therefore \dfrac{1.20}{R_S} = \dfrac{1000^{1/2}}{7000^{1/2}} \qquad R_S = 3.2$$

****

SAQ 2.5a

In Section 2.5 you worked out the effect of extra-column dispersion on the peak of an unretained solute, using a column with a plate number of 10 000. The extra-column dispersion will decrease the plate number that we actually observe for this column. The table below contains the retention volume, peak widths and plate number for an unretained solute on this column.

(a) Complete the table by calculating the corresponding values for columns with plate numbers of 5000 and 3000 respectively;

(b) What is the percentage reduction in the plate number of each column due to extra-column effects?

N (ideally)	10 000	5000	3000
V_R, μl		2908.5	2908.5	2908.5
w, μl	116		
w_T, μl	126		
N (actual)	8525		
% reduction in N			

Response

N (ideally)	10 000	5000	3000
V_R, μl	2908.5	2908.5	2908.5
w, μl	116	165	212
w_T, μl	126	172	218
N (actual)	8525	4573	2847
% reduction in N	15	8.5	5

In each case the plate numbers are calculated using Eq. (2.2a) and the actual peak with w_T. Thus, for the first column:

$$N = 16 \times (2908.5)^2 / (126)^2 = 8525$$

You can see that the effect of a given amount of extra-column dispersion is more serious the higher the efficiency of the column.

****

3. Solvent Delivery and Sample Injection

3.1. THE MOBILE PHASE RESERVOIR

The simplest reservoir is a 1 dm^3 glass bottle with the cap drilled to take a 1/8 in. diameter PTFE tube to carry the mobile phase from the reservoir to the pump. The liquid entering the pump should not contain any dust or other particulate matter, as this can interfere with the pumping action and can cause damage if it gets into the seals or valves. Such material can also collect on top of the column, causing irregular behaviour or maybe even blockages. The mobile phase is, therefore, filtered before it enters the pump. This can be done using a stainless steel filter element that is a push fit on to the end of the PTFE tube in the reservoir, or alternatively an in-line filter can be used. The pore size is normally 2 μm.

It is also important to remove dissolved air or suspended air bubbles. Pockets of air can collect in the pump or in other places, causing strange behaviour from the detector and irregular pumping action (at worst, the pumping action can be lost completely). Practical methods for degassing the mobile phase are dealt with in Section 9.2.

Figure 3.1a is a diagram of a commercially available mobile phase reservoir fitted with a tube for helium degassing, a push fit filter and a debubbler. The internal shape of the bottle allows almost all the mobile phase to be pumped without tilting the reservoir. The debubbler is a useful device fitted between the reservoir and the pump that traps out any air bubbles. The reservoir is coated on the outside with a UV absorbing plastic.

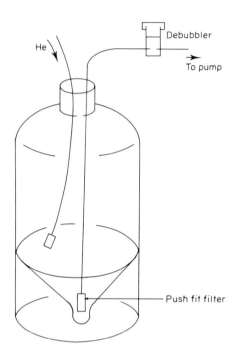

Fig. 3.1a. *Mobile phase reservoir*

3.2. PRESSURE, FLOW AND TEMPERATURE

Column inlet pressures in HPLC can be as much as 200 times atmospheric pressure, and HPLC columns are packed using much larger pressures (up to 700 times atmospheric). The SI unit of pressure is the Pascal (1 Pa = 1 Nm^{-2}); normal atmospheric pressure is about 10^5 Pa. Because it is convenient to express pressure using reasonably small numbers, experimental workers and instrument manufacturers report pressures in bar, or pounds per square inch (psi), or sometimes in kg cm^{-2}. The bar is defined by 1 bar = 10^5 Pa, so that 1 bar corresponds roughly to normal atmospheric pressure. You will need to be able to convert between these units. See if you can work out the conversion between bar and psi, given that 1 pound = 0.4536 kg, 1 in. = 2.54 cm, and g = 9.81 ms^{-2}.

1 psi = a force of 0.4536 × 9.81 N acting over an area of 0.0254^2 m^2.
1 psi = 0.4536 × 9.81/0.0254 = 6897 Pa
1 bar = 10^5/6897 = 14.5 psi.

The conversions for kg cm^{-2} are: 1 kg cm^{-2} = 0.981 bar = 14.2 psi. Although, as mentioned before, the column inlet pressure can be as much as 200 bar, most of the work in analytical HPLC is done using pressures between about 25 and 100 bar. The pressure developed will depend on the length of the column, the particle size of the stationary phase, and the viscosity and flow rate of the mobile phase. Because liquids are not very compressible there is not much energy stored in them at high pressures, and the pressures used in HPLC do not represent a hazard (precautions should be taken when packing columns, when the pressures used are much higher). The pressures above correspond to mobile phase flow rates of roughly 1–5 cm^3 min^{-1} through the column. For constant flow pumps (Section 3.3.2) the mobile phase flow rate can be set at the pump, although for some cheaper pumps the flow setting is not very reliable, and will not equal the flow through the column if there are leaks anywhere in the system. The mobile phase flow can be measured at the column outlet by collecting the solvent for a known time and weighing it, or measuring its volume.

Many commercial HPLC instruments provide a forced air oven which will control temperature with a stability of typically 0.1 °C from ambient temperature to 100 °C. Because of the use of flammable solvents, safety considerations are important, so the ovens are usually provided with a facility for nitrogen purging, and are designed to prevent the build up of solvent vapour in the event of a leak. If temperature control is used, it is important that the sample and the mobile phase are at the right temperature before being introduced to the column, so the mobile phase is normally passed through a preheating coil housed in the oven before it reaches the injection point.

Temperature control is important for the accurate measurement of retention data, and has to be used with refractometer detectors (Section 5.7). Increasing the temperature can increase the speed of the separation, especially in exclusion chromatography, and usually increases the efficiency of the column (though the gain in efficiency can be lost if the mobile phase is not properly equilibrated). Complicated separations can often be optimized by increasing the temperature, but this is done very much on a trial and error basis, and a lot of work in HPLC is still done at room temperature without temperature control.

3.3. PUMPS—GENERAL CONSIDERATIONS

The function of the pump in HPLC is to pass mobile phase through the column at a controlled flow rate. One class of pump (constant pressure pump) does this by applying a constant pressure to the mobile phase; the flow rate through the column then being determined by the flow resistance of the column and any other restrictions between the pump and the detector outlet.

Another type (constant flow pump) generates a given flow of liquid, so that the pressure developed depends on the flow resistance. The flow resistance of the system may change with time, owing to swelling or settling of the column packing, small changes in temperature, or build up of foreign particulate matter from samples, pump or injector. If a constant pressure pump is used the flow rate will change if the flow resistance changes, but for constant flow pumps changes in flow resistance are compensated for by a change of pressure. Since flow changes are undesirable, as they will cause retention data to lack precision and may cause an erratic baseline on the recorder, it is advisable not to use constant pressure pumps in HPLC instruments. However, they are suitable for packing columns, where small changes in flow do not matter.

In addition to being able to pump solvent at high pressure and constant flow, the pump should also have the following characteristics:

(*a*) The interior of the pump should not be attacked by any of the solvents that are to be used;

(*b*) A range of flow rates should be available, and it should be easy to change flow rate;

(*c*) The solvent flow should be non-pulsing;

(*d*) It should be easy to change from one solvent to another;

(*e*) The pump should be easy to dismantle and repair.

These points require some comment:

(*a*) All of the wetted parts of the pump should be made of inert materials of the sort described in Section 1.2. Even with these materials the pump may not like some solvent systems, for instance high concentrations of chloride ion or citrates will slowly corrode stainless steel.

(b) The pressures needed to achieve the flow rates required in analytical HPLC are not likely to exceed 150–200 bar. Most pumps are rated for much higher pressures than these (see Fig. 3.3a). At high pressure, the flow rate through the column may be less than the set flow, owing to the compressibility of the solvent or to small leaks.

(c) Some types of constant flow pump produce a pulsing flow of mobile phase. If the detector used is flow sensitive, the pulsing flow may produce baseline noise on the chromatogram, the recorder pen describing a fluctuating trace that tracks the motion of the pump piston. These flow variations can be reduced using a pulse damper, which is a dead volume placed between the pump and injector. Sometimes the dead volume of the column itself is enough to give satisfactory damping; if not, a small coil of tubing can be used between the pump and injector. However, large dead volumes between the pump and injector should be avoided (see below).

(d) The internal volume of the pump and of the plumbing between pump and injector should be kept as small as possible, and the system should not have any off-line recesses, otherwise changing the composition of the mobile phase will take a long time as the new phase displaces the old. An example of an off-line recess is the debubbler (Fig. 3.1a). This creates a recess that is not swept by the liquid flow. If the new mobile phase is immiscible with the original, then the changeover has to be done via a third solvent that is miscible with each.

(e) Even if great care is taken with the pump, the seals, rings and gaskets sometimes have to be replaced, so they should be easy to get at. Although chemically resistant, the seals and rings are relatively soft and are prone to wear. If the mobile phase has not been properly filtered, small particles can get trapped between the ball and the valve seat in a check valve (see Fig. 3.3b), causing the valve seat to wear. During the operation of the pump it is possible for solution to creep between the piston and the seal. This solution may evaporate (when the pump is idle), and if it contains dissolved solids (as in buffer solutions) a solid deposit will be left on the piston, causing damage when the pump is used again. Bacteria can sometimes grow in buffers; if such material is pumped it may block the column, or even cause blockages in the internal filters of the pump. Buffer solutions must therefore be carefully removed after use, by pumping water or a suitable solvent for several minutes, to flush the system.

Figure 3.3a lists some of the properties of three different types of pump (the operating principles of each are explained in the next section).

Model	Shimadzu LC-9A	ISCO LC 500	Stanstead A9512 LC
Type	constant flow twin reciprocating	constant flow syringe type	constant pressure pneumatic amplifier
Maximum output pressure, bar	392	250	500
Flow rate range, cm^3 min^{-1}	0.001–5	1.3×10^{-4}–3.34	up to 200
Capacity, cm^3	continuous pumping	375	continuous pumping

Fig. 3.3a.   *Operating properties of three different types of pump*

3.3.1. Constant Pressure Pumps

The earliest form of constant pressure pump in HPLC (the coil pump) used pressurized gas from a cylinder to drive mobile phase from a holding coil through the column. This type of pump was used in some of the older HPLC instruments, but is now only of historical interest. If you want to know any more about it, there are details in most text books.

The operating principle of the pneumatic amplifier pump is shown in Fig. 3.3b.

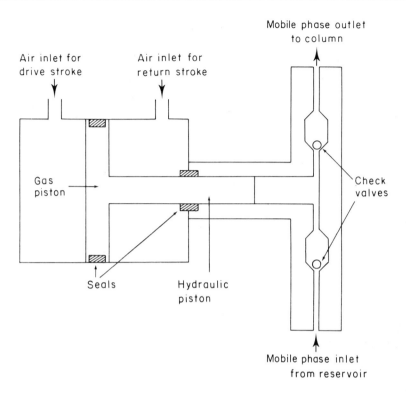

Fig. 3.3b. *Pneumatic amplifier pump*

Air from a cylinder at pressures up to about 10 bar (150 psi) is applied to a gas piston that has a relatively large surface area. The gas piston is attached to a hydraulic piston that has a smaller surface area. The pressure applied to the liquid = gas pressure × area of gas piston/area of hydraulic piston. With 10 bar inlet pressure and a 50:1 area ratio, the hydraulic pressure obtained is 500 bar (7500 psi). On the drive stroke, the outlet valve on the pump head is open to the column and the inlet valve is closed to the mobile phase reservoir. At the end of the drive stroke, the air in the chamber is vented and air enters on the other side of the gas piston to start the return stroke. On the return stroke the outlet valve closes, the inlet valve opens and the pump head refills with mobile phase. The pump can be started and stopped by operation of a valve fitted between the cylinder regulator and the pump.

Compared to syringe type or reciprocating pumps, pneumatic amplifier pumps are very cheap. They tend to be rather difficult to get into for repairs, and some types are very noisy in operation. Because they do not provide a constant flow of mobile phase they are not used much in analytical HPLC. They can, however, operate at high pressures and flow rates and so are used mainly for packing columns, where high pressures are needed and variations in the flow rate through the column do not matter.

3.3.2.   Constant Flow Pumps

Two types of constant flow pump have been used in HPLC. Figure 3.3c shows a syringe type pump.

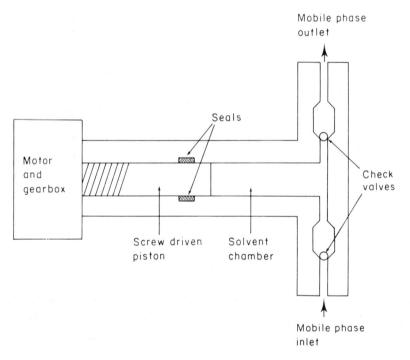

Fig. 3.3c.   *Syringe pump*

Mobile phase is displaced from a chamber by using a variable speed stepper motor to turn a screw which drives a piston. The chamber has a volume of 200–500 cm^3. The flow is pulseless and can be varied by changing the motor speed. The mobile phase capacity is limited to the volume of the solvent chamber. Although this is fairly large, so that many chromatograms can be run before the chamber has to be refilled, a lot of solvent is wasted in flushing out the pump when a change is required.

The type of pump used in most instruments is the reciprocating pump, shown in Fig. 3.3d.

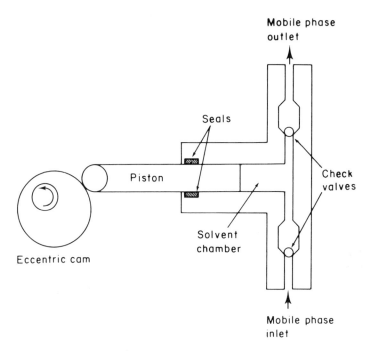

Fig. 3.3d. *Reciprocating pump*

The piston is driven in and out of a solvent chamber by an eccentric cam or gear. On the forward stroke, the inlet check valve closes, the outlet check valve opens, and mobile phase is pumped to the column; on the return stroke the outlet valve closes and the chamber is refilled. Unlike syringe pumps, reciprocating pumps have an unlimited capacity, and their internal volume can be made very small, from 10–100 μl. The flow rate can be varied by changing the length of stroke of the piston or the speed of the motor. Access to the valves and seals is usually fairly straightforward.

In the single headed reciprocating pump shown in the figure, the mobile phase is only being delivered to the column for half the time that the pump is in operation, and during the drive stroke of the piston the flow rate is not constant (because the speed of the piston changes with time). The output of the pump is shown in Fig. 3.3e(i). Use of a twin headed pump with the two heads operated 180° out of phase (so that while one head is pumping the other is refilling) produces the output shown in (ii).

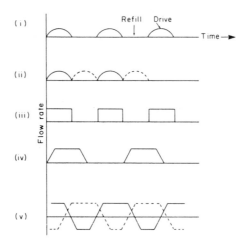

Fig. 3.3e. *Output from reciprocating pumps*

- (*i*) single headed pump
- (*ii*) twin headed pump, heads 180° out of phase
- (*iii*) single headed pump with constant speed (ideal)
- (*iv*) single headed pump with constant piston speed (in practice)
- (*v*) twin headed pump, heads 180° out of phase, with different constant speeds on the drive and refill strokes

Modern twin headed pumps use two pistons driven by a shaped cam or gear that produces a constant piston speed. Ideally, the output of one such head should be as shown in Fig. 3.3e(iii) operation of two heads 180° out of phase producing a pulseless flow. In practice, on each drive stroke the change in flow rate is not instantaneous, giving an output shown in (iv). To overcome this, the driving cam is arranged to make the piston travel faster on the refill than on the drive stroke, producing an output shown in (v). The flow rate, which is the sum of the output of both heads, is constant.

Another way of reducing flow noise with single headed pumps is to use a rapid stroke rate (one model uses 23 strokes s^{-1} so that the detector cannot react rapidly enough to sense the flow changes. Many pumps also use feedback flow control, where the flow rate is measured downstream of the pump; any difference between the measured and the set flow actuates a change of motor speed so as to reduce the flow difference to zero.

SAQ 3.3a

Your company has decided to manufacture an HPLC pump and they are trying to decide what materials to build it with. To a large extent this decision is influenced by the performance criteria that are set by the intended function of the unit. Write a brief list showing what you think are the important criteria.

SAQ 3.3b   The potential construction materials for the pump in
                 the preceding SAQ fall into four broad classes:

    (*i*) stainless steel;

    (*ii*) polymers;

    (*iii*) alloys;

    (*iv*) ceramics.

    Very briefly, describe how the properties of these
    materials fit the performance criteria in the previous
    list.

SAQ 3.3c

A number of manufacturers produce 'metal free' HPLC systems for trace level work, to avoid contamination of the mobile phase with traces of iron and other elements from the pump and other metal components. Other manufacturers argue that if the pump is properly passivated the metal contamination arises primarily from impurities in the reagents that are used to prepare the mobile phase.

Suppose you are determining transition metal cations using PAR derivatization (Section 7.6.5). The reagent is made up in a solution containing 3 mol dm^{-3} NH$_4$OH and 1 mol dm^{-3} CH$_3$COOH. The stated maximum level of iron in these two reagents is $2 \times 10^{-5}\%$ and $10^{-6}\%$, respectively. What is the maximum concentration of iron (ppb) that the use of this solution will produce?

3.4. GRADIENT FORMERS

Figure 3.4a (i)–(iii) shows block diagrams of three types of gradient former. At low pressure, gradients can be formed from solvents A and B by metering controlled amounts of A into B in a mixing chamber before the high pressure pump (i). Alternatively, the composition in the low pressuring mixing chamber can be controlled by using time proportioning valves as in (ii). This requires microprocessor control of the valves, which have to be

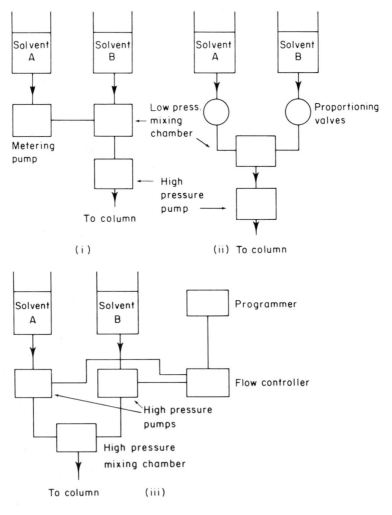

Fig. 3.4a. *Gradient formers*

switched very rapidly and accurately. In (iii), the separate solvents can be pumped with two high pressure pumps into a high pressure mixing chamber. The type of gradient formed is controlled by programming the delivery of each pump, but the use of two pumps is a very expensive means of forming solvent gradients.

3.5. SAMPLE INJECTION

The earliest injection method for HPLC used a technique borrowed from GC, in which a microlitre syringe was used to inject the sample through a self-sealing rubber septum held in an injection unit at the top of the column. In another method (stopped flow) the flow of mobile phase through the column was halted and, when the column reached ambient pressure, the top of the column was opened (e.g. using a ball valve at the column inlet) and sample introduced at the top of the packing. These two methods are quite cheap and are still used occasionally in home-made instruments. If you would like to read any more about them, there are good accounts in the textbooks by Knox and by Meyer (see Bibliography).

Commercial chromatographs use valves for sample injection. Although these are expensive, they are easy to use, give good precision and are easily adapted for automatic injection. With these devices, sample is first transferred at atmospheric pressure from a syringe into a sample loop (sample container). Turning the valve from load to inject position connects the sample loop into the high pressure mobile phase stream, whereby the contents of the sample loop are transferred on to the column.

Fig. 3.5a shows the flow paths for the very popular Rheodyne 7125 valve. Sample from a microlitre syringe is loaded into the needle port (4), filling the sample loop, which is a small piece of stainless steel tube connected between ports 1 and 4. Any excess goes to waste from port 6. On turning to 'inject', the loop contents are flushed on to the column from port 3.

A variety of loop volumes is available, commonly 10–50 μl. For smaller volumes than this, the loop is an engraved slot in the body of the valve, and four ports are used to carry the mobile phase in and out.

The stated volume of the loop is not accurate, because loops are made by cutting tubing to a specified length, not by volumetric calibration. The

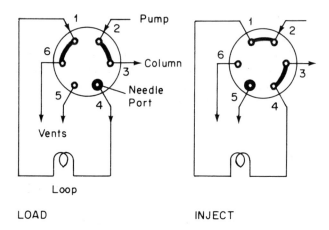

Fig. 3.5a. *Flow diagram for Rheodyne 7125 valve*

tolerance on the internal diameter of the tubing can cause quite large errors for small volume loops. It is possible to calibrate the loop, but it is not often necessary to know the exact volume injected. We do, however, often have to make precise injections, and the way that the valve is used can affect the precision. If you change injection volume infrequently and can afford to waste a little sample, the loop should be overfilled. This is done by loading a 2-5 fold excess of sample. The reason we need to use this much is because of the flow profile of sample in the loop, sample flowing along the centre axis of the loop at twice the average velocity and reaching the end of the loop when only half a loop volume has been loaded (see Fig. 2.5b). Complete filling of the loop in this way gives very good precision, but the volume injected can only be changed by changing the loop.

If you frequently change injection volume, or if you can't afford to waste any sample, the loop can be partially filled using a syringe, in which case the volume injected is the volume dispensed from the syringe, provided it is no more than half the loop volume. The precision this time will be the precision of the syringe, which is not as good as that of the valve. After injection, a small amount of sample will remain in the needle port of the valve. This should be flushed out with mobile phase (with the valve in inject position) before the next injection. If the mobile phase contains dissolved salts, the flow passages should be flushed out with water before the valve is left, otherwise crystals may form inside the valve.

3.6. SUMMARY

Practical methods for the delivery of mobile phase are described, including the operating principles of three types of pump together with their advantages and limitations. Techniques for the production of gradients and for the introduction of samples are considered.

Learning Objectives

Now that you have completed Chapter 3 you should be able to:

• Specify typical flow rates and pressures used in HPLC;

• Identify the characteristics required of a pump for HPLC;

• Distinguish between constant pressure and constant flow types and understand the working principles of each type;

• Outline simple methods for the production of gradients, and explain the operation of an injection valve.

References

1. Veronika R. Meyer, *Practical High Performance Liquid Chromatography*, Wiley, 1988, Chapter 3.

2. J.H. Knox, Ed. *High Performance Liquid Chromatography*, Edinburgh University Press, 1982, Chapter 9.

3. L. Berry and B.L. Karger, *Analytical Chemistry* 1973, 45, 819a–828A.

4. J.W. Dolan, *LC–GC International* 1991, 4(6), 10–14; 1991, 4(5), 20–22.

SAQS AND RESPONSES

SAQ 3.3a   Your company has decided to manufacture an HPLC pump and they are trying to decide what materials to build it with. To a large extent this decision is influenced by the performance criteria that are set by the intended function of the unit. Write a brief list showing what you think are the important criteria.

Response

Important performance criteria are:

(i) Materials should be able to withstand pressures in excess of 7000 psi;

(ii) Precise flow delivery;

(iii) Inertness–non-reactive to samples and eluents;

(iv) Impermeability–prevention of contamination by ingress of atmosphere or light.

(v) Reliability–leak free connections and long service life.

(vi) Low raw material and machining costs.

(vii) Good thermal stability (range about 0–100 °C).

****

SAQ 3.3b

> The potential construction materials for the pump in the preceding SAQ fall into four broad classes:
>
> (*i*) stainless steel;
>
> (*ii*) polymers;
>
> (*iii*) alloys;
>
> (*iv*) ceramics.
>
> Very briefly, describe how the properties of these materials fit the performance criteria in the previous list.

Response

Ceramics are inert, impermeable, pressure resistant and thermally stable but are brittle materials, and the machining and forming of ceramics presents serious problems for high precision devices. Alloys have most of the desirable properties required, with the exception of cost, which is prohibitive compared with stainless steel or plastics.

Both stainless steel and some polymers are reliable construction materials. Threads and fittings made from polymers will be more prone to cross-threading, stripping and leaks. Stainless steel fittings are harder to destroy, though it can be done with a little determination. Stainless steel will be superior in respect of thermal stability and flow precision. Both stainless steel and suitable polymers show good chemical resistance to most eluents and samples. Polymers may be permeable to both gases and light.

****

SAQ 3.3c

A number of manufacturers produce 'metal free' HPLC systems for trace level work, to avoid contamination of the mobile phase with traces of iron and other elements from the pump and other metal components. Other manufacturers argue that if the pump is properly passivated the metal contamination arises primarily from impurities in the reagents that are used to prepare the mobile phase.

Suppose you are determining transition metal cations using PAR derivatization (Section 7.6.5). The reagent is made up in a solution containing 3 mol dm^{-3} NH_4OH and 1 mol dm^{-3} CH_3COOH. The stated maximum level of iron in these two reagents is $2 \times 10^{-5}\%$ and $10^{-6}\%$, respectively. What is the maximum concentration of iron (ppb) that the use of this solution will produce?

Response

3 mol dm^{-3} NH_4OH = 105 g dm^{-3} or 2.1×10^{-5} g dm^{-3} Fe

1 mol dm^{-3} CH_3COOH = 60 g dm^{-3} or 6×10^{-7} g dm^{-3} Fe

Total = 2.16×10^{-5} g dm^{-3} Fe or 21.6 ppb.

****

4. Columns

4.1. DIMENSIONS AND FITTINGS

The columns most commonly used at the moment are made with 316 grade stainless steel (a Cr–Ni–Mo steel, relatively inert to chemical corrosion). The inside of the stainless steel tube should be as smooth as possible, so the tubes are precision drilled or electropolished after manufacture. Common dimensions are 6.35 mm ($\frac{1}{4}$ in.) external diameter, 4.6 mm internal diameter and up to 25 cm long. Most manufacturers offer a range of lengths and diameters, e.g. lengths of 10, 12.5 or 15 cm and internal diameters of 3, 4.6, 6.2 or 9 mm. The columns can be packed with 10, 5, 4 or 3 μm diameter particles.

At the top of the column, there is a distributor for directing the injected sample to the centre of the column and then a stainless steel gauze or frit on top of the packing. At the lower end there is another frit to retain the packing, and then, for the 4.6 mm type, a reducing union and a short length of 0.25 mm (0.01 in.) i.d. tubing to connect the column to the detector. (I hope you don't find the mixture of units too confusing; manufacturers tend to give external diameters and fitting sizes in inches, internal diameters in mm and column lengths in cm!). The appearance of a typical column in shown in Fig. 4.1a.

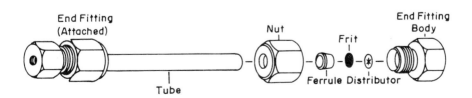

Fig. 4.1a. *A typical conventional column*

Materials other than stainless steel that are used for columns include glass, glass lined steel tube and polyethene or other inert plastics.

At each end of the column is a reducing union. Conventional reducing unions have rather a large dead volume, so they are bored out to produce a zero dead volume (ZDV) union in which both the metal column and the external tubing are butted up directly against the stainless steel frit. There is some evidence that the abrupt change in diameter in the ZDV fitting where the two tubes meet can cause some loss of efficiency, so many columns use the more expensive low dead volume (LDV) fitting in which the frit at the end of the column is followed by a shallow distributive cone leading to the external tube. The dead volume of the LDV type is very small, often about 0.1 μl. The three types are shown in Fig. 4.1b.

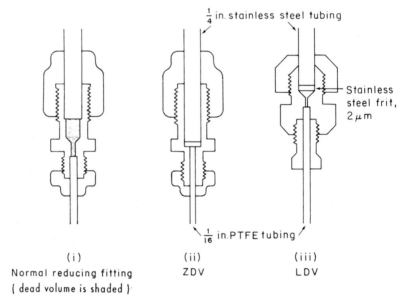

Fig. 4.1b. *Column outlet fittings*

The 4.6 mm column is connected at the upper end to an injection valve using a ZDV or LDV reducing union and a short length of stainless steel or high pressure plastic tubing.

Columns are sold by manufacturers with a $\frac{1}{4}$–$\frac{1}{16}$ ZDV or LDV reducing union at the outlet and a $\frac{1}{4}$ inch nut and cap or a reducing union at the inlet. The column will have been tested and a test chromatogram will be supplied with it. The test chromatogram is, of course, designed to show the column in the best possible light, and the column may not behave as well in another instrument, or with another sample.

Needless to say, the fittings and thread sizes used by different manufacturers are not all compatible. This can be a real headache in practice, so it is worth looking at the problems in a little more detail. Figure 4.1c(i) shows the appearance of a male nut and ferrule after having been tightened into a union. The dimension X that is obtained is different for different manufacturers. Part (ii) shows typical values of X (not specifications) for unions made by different manufacturers. Where the dimension is the same, as in (iii), the fittings can be mixed. Parts (v) and (vi) of the figure show what happens if we use the incompatible fittings that are listed in part (iv).

∏   What would be the consequences of doing this?

Either the dimension X is too long so the ferrule does not seat properly and the fitting leaks, or else X is too short, giving undesirable dead volume in the fitting.

The most common thread sizes used in HPLC fittings are $\frac{1}{4}$-28, 10-32 and M6. For the English type threads, the first number is the diameter of the threaded portion of the fitting (not the tubing) and the second number gives the number of threads per inch. However, thread sizes smaller than $\frac{1}{4}$ in. are described by a number (from 1–12, corresponding to diameters of 0.073–0.216 in.). A 10-32 thread is 0.190 in. in diameter and has 32 threads per in. An M6 fitting has one thread per mm (standard metric) and a diameter of 6 mm. Male nuts with similar thread sizes cannot be interchanged if the length of the threads differs.

A welcome development has been the advent of reliable high pressure plastic fittings and tubing. Most chromatography suppliers now offer a range of plastic fittings that can be tightened to be leak free, by hand. These are commonly made of Kel-F (PCFE) or Peek (a polyketone). Peek has excellent chemical resistance to most organic and inorganic liquids (exceptions are concentrated nitric and sulphuric acids and tetrahydrofuran) and the tubing can be used at pressures up to 414 bar (6000 psi) for the smaller diameters.

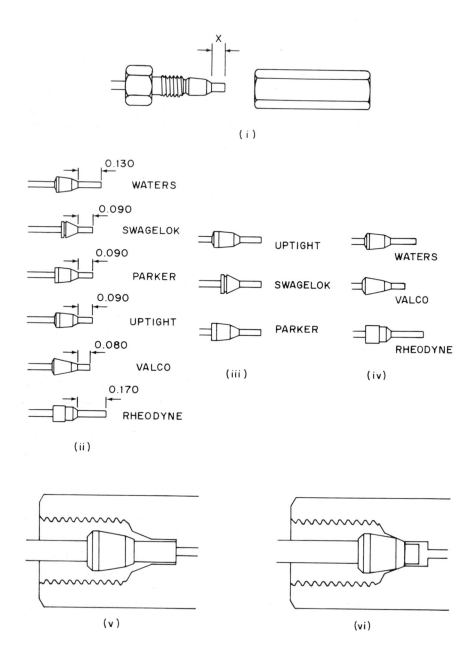

Fig. 4.1c. *Fittings from different manufacturers*

4.2. CARTRIDGE SYSTEMS

A number of manufacturers produce columns in various materials that use a 'cartridge' system, in which the column is supplied as a cartridge and the end fittings are either tightened on by hand or are contained in a separate cartridge holder, the column being fixed in the holder in some systems simply by operating a compression lever which seals the column in. These columns are thus much easier to change than conventional columns that are terminated with compression fittings. Because they are sold without end fittings, cartridge-type columns are also cheaper than conventional columns, although the initial outlay on the cartridge holder has to be considered as well.

4.3. AXIAL AND RADIAL COMPRESSION

A source of dispersion with stainless steel columns is in the wall regions, where the presence of interparticle voids is likely; this leads to mixing and thus band spreading.

To reduce these effects, the Waters 'radial compression system' uses a polyethene cartridge column which is inserted into a radial compression module. The flexible walls of the column are subjected to hydrostatic pressure which moulds the walls of the column around the packing and compresses the column bed, thus producing an improvement in the packed bed structure, increased efficiency and longer column life. The principle of the method is shown in Fig. 4.3a.

Axial compression is a technique for removing voids at the top of the column which can be caused, for instance, by the gradual settling or dissolution of the column packing. This produces unwanted dead space at the column inlet and can lead to serious loss of efficiency, as shown in Fig. 9.3b. Axial compression columns have a movable piston at the top of the column which can be tightened down by hand to remove any voids that have been formed.

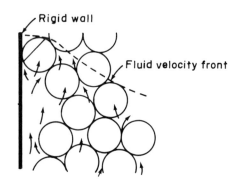

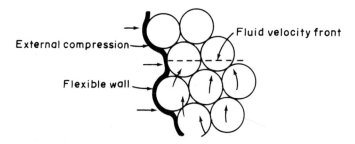

Fig. 4.3a. *Radial compression*

4.4. SUMMARY

HPLC column materials, dimensions and end fittings are discussed, together with the techniques of axial and radial compression.

Learning Objectives

Now that you have completed Chapter 4 you should be able to:

- Identify typical materials and dimensions of HPLC columns;

- Recognize that compatibility problems can occur with different column fittings;

- Understand the principles of axial and radial compression.

5. Detectors

5.1. INTRODUCTION

The function of the detector in HPLC is to monitor the mobile phase emerging from the column. The output of the detector is an electrical signal that is proportional to some property of the mobile phase and/or the solutes. Refractive index, for example, is a property of both the solutes and the mobile phase. A detector that measures such a property is called a *bulk property* detector. Alternatively, if the property is possessed essentially by the solute, such as absorption of UV/visible radiation or electrochemical activity, the detector is called a *solute property* detector. Quite a large number of devices, some of them rather complicated and temperamental, have been used as HPLC detectors, but only a few have become generally used, and we will examine five such types. Before doing this, it is helpful to have an idea of the sort of characteristics that are required of an ideal detector.

Π   What do you think they are? (there are about seven important ones).

The important characteristics are:

(*a*) Sensitivity.

(*b*) Linearity.

(*c*) Universal or selective response.

(*d*) Predictable response, unaffected by changes in conditions.

(*e*) Low dead volume.

(*f*) Nondestructive.

(*g*) Cheap, reliable and easy to use.

No detector possesses all of these, and unfortunately the properties are not all independent of one another, so that improving a detector in one respect may make it worse in another.

(*a*)   Sensitivity is the ratio of output to input, so that we want a large detector signal for a small amount of solute. Any detector will suffer from instrumental, principally electronic, noise, the amplitude of which will determine the minimum amount of solute that we can detect, because as we decrease the amount of our solute, eventually the detector signal will become so small that it cannot be distinguished from the noise. The sensitivity of detectors is often given as a noise equivalent concentration, C_N, which means the concentration of solute that produces a signal equal to the detector noise level. C_N will depend on the nature of the solute that is used to make the test.

(*b*)   A linear detector has a response that is directly proportional to the amount or concentration of solute. The linear range of the detector is that concentration range over which this proportionality is obeyed. It is still possible to use a detector if it is not linear, but quantitative work becomes much more difficult.

(*c*) and (*d*) A universal detector will sense everything in the sample, whereas a selective detector will sense only certain components. In analytical work we need both types. Ideally, our detector would have the same very high sensitivity for all solutes, be capable of operating universally or selectively and have a response that did not depend on the operating conditions, but this is asking far too much of it. The best we can hope for is that we can predict how the response of the detector will change for different chemical types, and that the response does not change too much if there are small changes in the operating conditions (e.g. column temperature or flow rate).

(*e*)   Dead volume in the detector adds to extra-column dispersion, so it must be kept to a minimum. This includes the cell volume of the detector itself, and also the length and bore of any tubing associated with it. For spectrometric detectors a reduction in the cell volume is likely to lead to a loss of sensitivity.

Some of these characteristics are listed for different detectors in Fig. 5.1a.

Type	Response	Noise level	C_N g cm^{-3}	Linear range	Flow cell volume μl
UV–visible absorption	selective	10^{-4} a.u.	10^{-8}	10^4–10^5	1–8
Fluorescence	selective	10^{-7} a.u.	10^{-12}	10^3–10^4	8–25
Conductivity	selective	10^{-2} μS cm^{-1}	10^{-7}	10^3–10^4	1–5
Amperometric	selective	0.1 nA	10^{-10}	10^4–10^5	0.5–5
Refractive index	universal*	10^{-7} r.i.u.	10^{-6}	10^3–10^4	5–15

* There must be a difference between the refractive index of the solutes and that of the mobile phase.

Fig. 5.1a. *Characteristics of detectors used in HPLC; a.u. = absorbance units; r.i.u. = refractive index units*

Noise levels will be different for different models of the same type of detector, and for a given model will depend very much on how the detector is used. The noise equivalent concentration refers to a solute with favourable properties, and may be very much higher for other solutes.

5.2. DETECTION LIMITS

If we want to know the smallest amount that we can measure in a particular separation, the most frequently used indication of this is the detection limit, which tells us the minimum amount of a species that can be reliably seen on the chromatogram. Usually detection limits are measured with an individual species in a standard solution—in a real sample the detection limit for this species will usually be higher (i.e. worse) due to baseline disturbances or interference from the matrix or other species.

Experimentally, the detection limit is the amount of substance that will give a peak whose height is some multiple of the baseline noise. Multiples of 2, 3 or 5 are commonly used (this is the signal to noise ratio S/N); the lower the multiple used the better the detection limit appears. Detection limits are often quoted by manufacturers as a concentration (e.g. in ppm or ppb). Unfortunately, this does not mean very much unless we also know the S/N that was used and the injection volume, so that we can work out the mass of substance that was placed on the column.

To evaluate a detection limit we first have to measure a section of typical baseline noise. We have to decide what 'typical' is. Note that the noise will often contain two components, a small amplitude high frequency component and also low frequency bumps or drift; both of these must be allowed for. Generally one of the tricks used to obtain unrealistically low detection limits is to consider only the high frequency component. We then measure the height of the peak (with the same units as were used for the noise) and calculate the amount of substance that would produce a peak with 2, 3 or 5 times the height of the baseline noise.

SAQ 5.2a

Estimate the detection limit for Cl^- from the chromatogram shown in Fig. 5.2a. Use 2 × S/N and express the result both as concentration and as mass of Cl^-.

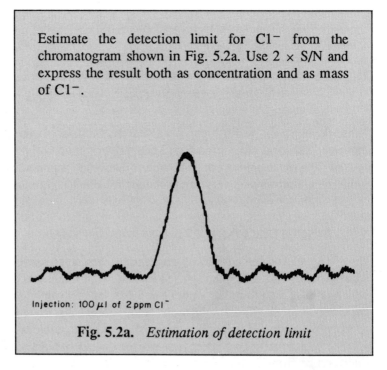

Injection: 100 μl of 2 ppm Cl^-

Fig. 5.2a. *Estimation of detection limit*

SAQ 5.2a

5.3. UV ABSORBANCE DETECTORS

These are by far the most popular detectors in HPLC. The principle is that the mobile phase from the column is passed through a small flow cell held in the radiation beam of a UV/visible photometer or spectrophotometer. These detectors are selective in the sense that they will detect only those solutes that absorb UV (or visible) radiation. Such solutes include alkenes, aromatics and compounds having multiple bonds between C and O, N or S. The mobile phase we use, on the other hand, should absorb little or no radiation.

Absorption of radiation by solutes as a function of concentration, c, is described by the Beer–Lambert law:

$$A = ecb \tag{5.3a}$$

where A = absorbance, b = path length of the cell and e = molar absorptivity, which is a constant for a given solute and wavelength.

The magnitude and the units of the absorptivity in Eq. 5.3a will depend on the units of c and b. In SI units, with c in mol m^{-3} and b in m, e is in mol^{-1} m^2, but it is common practice to measure c in mol dm^{-3} and b in cm, when e will be in dm^3 mol^{-1} cm^{-1}.

Strictly, the Beer–Lambert law applies only to monochromatic radiation. However, the detector system does not provide truly monochromatic radiation, but rather a narrow band of wavelengths centred around the selected wavelength. If we examine the law for a solute at a wavelength where the absorbance is changing rapidly, then the different wavelengths comprising the band may be absorbed by quite different amounts and the law may not be obeyed. In theory, we want to operate our detector in flat regions of the spectrum (maxima, minima or shoulders). In practice, if we are detecting more than one component and our detector is not programmable, it is usually not possible to do this (see for example Fig. 8.4d). Whether or not the law is obeyed in steep regions of the spectrum depends on the quality of the detector (for modern UV/visible detectors this is not normally a problem).

Both fixed and variable wavelength UV/visible detectors are available. The variable types use a deuterium and/or a tungsten filament lamp as the radiation source, and can operate between about 190 and 700 nm. They will have a number of switched sensitivities (absorbance ranges) measured in 'a.u.f.s.', which means absorbance units corresponding to full scale deflection on the recorder. Fixed wavelength detectors normally operate at 254 nm or 280 nm, but other wavelengths are possible. Figure 5.3a shows part of the specification of two modern variable wavelength detectors.

Detector	Radiation source	Wavelength range, nm	Bandwidth, nm	Absorbance ranges a.u.f.s.	Noise a.u.
Cecil CE 1200	Deuterium lamp (tungsten lamp accessory)	190–400 380–600	10	0–2 (10 ranges)	10^{-5}
Waters 484	Deuterium lamp	190–600	8	0–2	1.5×10^{-5}

Fig. 5.3a. *Part of the specification of two variable wavelength detectors*

∏ Suppose you are using a UV detector with a noise level of 10^{-4} a.u. The detector is linear from the noise level to $A = 1$, and a flow cell with a path length of 10 mm is used.

(a) What is the linear range of the detector?

(b) Use Eq. 5.3a to calculate C_N (in g cm^{-3}) for a solute with M_r = 100 and absorptivity = 1000 dm^3 mol^{-1} cm^{-1}.

(a) 10^4

(b) If c = the concentration of solute that produces an absorbance of 10^{-4}, then $10^{-4} = 1000 \times 1 \times c$.

$$\therefore c = 10^{-7} \text{ mol dm}^{-3}$$

$$= 10^{-5} \text{ g dm}^{-3}$$

$$= 10^{-8} \text{ g cm}^{-3}$$

This would be lower for another compound that had a higher absorptivity.

Fig. 5.3b is a diagram of a simple type of UV flow cell. The cell has a 1 mm internal diameter and the optical path is 10 mm giving it an internal volume of just under 8 μl. Modern instruments use a 'cassette' type flow cell, which plugs into a holder in the detector; in older detectors there are arrangements for horizontal and vertical adjustment of the position of the flow cell in the radiation beam. More complicated flow cell designs attempt to reduce flow disturbances, which can be caused by changes in the refractive index of the eluent, for instance if the solute has been dissolved in a solvent other than the mobile phase. Fig. 2.5b shows that the material will appear first in the centre of the flow cell and last at the walls, forming what amounts to a moving lens of liquid in the cell. This distorts the radiation beam, and may either increase or decrease the amount of radiation falling on the sensor in the instrument. Such changes are often seen as a differential peak on the chromatogram at about the time expected for an unretained solute.

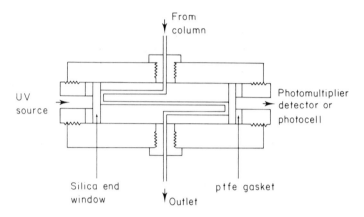

Fig. 5.3b. *Flow cell for UV/visible absorbance detector*

Another common problem is that the detector registers very high or off scale absorbance readings all the time, i.e. the UV radiation is being absorbed strongly when it should not be.

Π   Can you think of any reasons why this might happen?

The possible causes are:

(a) The mobile phase contains some UV absorbing component. This can be checked by measuring the absorbance of the mobile phase using another spectrophotometer, but make sure that you take the sample directly from the mobile phase reservoir. It is not unknown for the mobile phase to have been made up incorrectly, and if you make up a fresh sample you might get it right!

(b) There are large air bubbles in the flow cell. These can sometimes be removed by pumping at a high flow rate or by disconnecting the detector from the column and passing solvent rapidly through the flow cell using a syringe. This can be done with a syringe of 10–20 cm^3 capacity with a 1/16 union fitted to the end of the needle.

(c) The flow cell may be leaking so that there are drops of solvent on the outside of the end windows, or the end windows may be dirty, or the cell may not be properly aligned in the instrument. The alignment is easily checked, but only dismantle and clean the flow cell as a last resort; some types are quite difficult to reassemble. Faults in the detector can be checked by seeing if you can obtain zero absorbance with the flow cell removed; if you can, the detector is probably all right.

SAQ 5.3a

Fig. 5.3c shows the UV spectra of azobenzene (Az, concentration 3.73×10^{-3} g dm^{-3}) and phenanthrene (P, 3.23×10^{-3} g dm^{-3}) both recorded in *iso*-octane on a standard UV/visible spectrophotometer. The wavelength drive on the instrument was 10 nm cm^{-1} and the absorbance range was 2 a.u.f.s. Measurements were made against *iso*-octane using 10 mm cells.

What wavelength would you choose:

(*i*) To detect Az without detecting P;

(*ii*) To detect P without detecting Az;

(*iii*) To detect both of them;

(*iv*) To detect Az at maximum sensitivity?

$\longrightarrow$

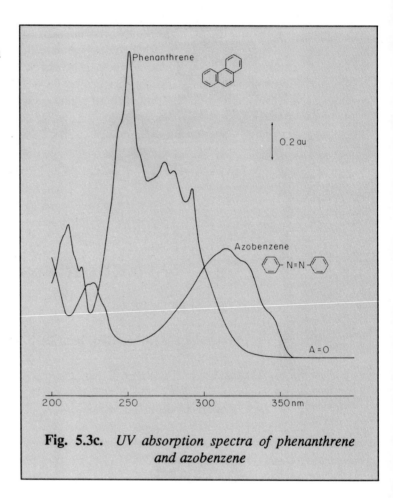

Fig. 5.3c. *UV absorption spectra of phenanthrene and azobenzene*

SAQ 5.3b

Infrared absorption detectors are available for HPLC, although they have never become very popular. From what you know about IR spectrometry and what you have read so far about HPLC detectors, see if you can decide whether the following statements are true or false.

(*i*) An infrared spectrum provides more structural information about a compound than does a UV spectrum;

(*ii*) An IR detector would be more sensitive than a UV detector;

(*iii*) An IR detector could not be used with solvent mixtures containing water;

(*iv*) An IR detector could be used as a selective detector or a universal detector by changing the wavelength used.

5.4. THE PHOTODIODE ARRAY DETECTOR (PDA)

In a conventional UV/visible spectrophotometer, polychromatic radiation is passed through the sample and is then focussed on to the entrance slit of a monochromator, which passes a narrow band of wavelengths to the detector. The absorbance of a sample is found by comparing the intensity of radiation reaching the detector without the sample (the blank) and after passing through the sample. To measure absorbance at different wavelengths, or to obtain spectra, the wavelength is changed by slowly rotating a prism or grating in the monochromator. The presence of moving parts in the instrument introduces errors (e.g. due to mechanical backlash), and also the process of recording a spectrum is relatively slow, since the data are acquired serially.

In the PDA, polychromatic radiation, after passing through the sample, is dispersed by a fixed grating and then falls on to an array of photodiodes. Each diode measures a narrow band of wavelengths in the spectrum, thus the PDA has parallel data acquisition, all points in the spectrum being measured simultaneously. This system has a number of advantages, some of which are as follows.

(*a*) Because of the parallel data acquisition, processing and storage of a spectrum can be done typically in about 0.5 s, compared with several minutes for a conventional instrument.

(*b*) As there are no moving parts to wear out, wavelength resetting errors are reduced and the instrument is likely to require less maintenance than does a conventional spectrophotometer.

(*c*) The ability to make multiwavelength measurements and the speed of data acquisition mean that various signal averaging techniques can be used to reduce noise and improve sensitivity.

When used as a detector for HPLC, the PDA, although more expensive than conventional UV detectors, has a number of significant advantages also. The spectrum of each peak in the chromatogram can be stored and subsequently compared with standard spectra, which facilitates the identification of peaks. The optimum wavelength for single wavelength detection can easily be found, or wavelength changes can be programmed to occur at different points in the chromatogram, either to provide maximum sensitivity for peaks, or to edit out unwanted peaks, or both. The instrument can provide a contour plot, showing the relationship between absorbance, wavelength and time. This can often be used for the detection and identification of otherwise unsuspected impurities in the sample.

Fig. 5.4a shows the spectra of two substances X and Y, together with a chromatogram and contour plot. The lines on the contour plot join points of equal absorbance (in commercial instruments the absorbance values are shown by using different colours for different absorbance ranges). The optimum wavelength for single wavelength detection would be 260 nm,

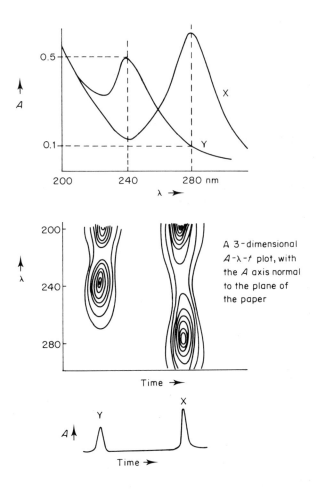

A 3-dimensional A-λ-t plot, with the A axis normal to the plane of the paper

Fig. 5.4a. *Absorption spectra and contour diagram*

alternatively both peaks could be determined at maximum sensitivity by starting the chromatogram at 240 nm and changing to 280 nm after Y has eluted.

The contour diagram for the peak shown in Fig. 5.4b(i) suggests an impurity at the leading edge of the peak. This can be investigated further by displaying the spectra obtained at different points across the peak, e.g. at points a, b and c as in (ii). The spectra are normalized (i.e. adjusted to allow for the different concentrations present at the three points) and overlaid. If the peak is not chromatographically pure then the spectra will not overlay property, as in (iii). Remember that 'pure' in this context means only that no other substance can be detected; there may well be a coeluting peak which has no UV activity and which therefore cannot be detected.

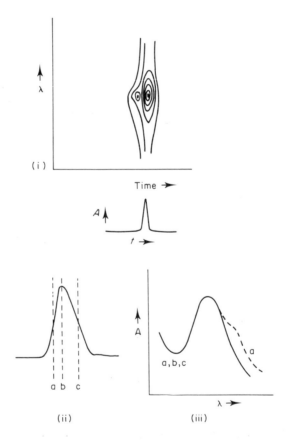

Fig. 5.4b.   *Contour diagram and overlaid spectra for an impure peak*

Since UV spectra are generally broad and featureless, it is sometimes difficult to decide if the spectra overlay properly or not, especially if the components concerned have similar spectra or if the concentration of one of them is small. In such cases it may be helpful to present the spectra as derivatives, especially second derivatives. Fig. 5.4c shows how

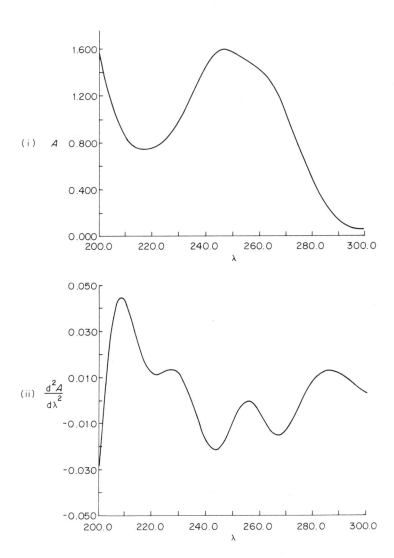

Fig. 5.4c. *Zero order (i) and second derivative (ii) UV spectra. Sample: 40 ppm thiamine hydrochloride*

an essentially nondescript zero order spectrum can be enhanced by presenting it as a derivative. The use of derivatives, however, always increases noise (which has low amplitude but high frequency) at the expense of the spectrum (high amplitude, low frequency). If the spectrum is noisy to start with, differentiating it may produce rubbish.

Another way of investigating peak purity is to determine absorbance ratios. Returning to Fig. 5.4a, the ratio A_{280}/A_{240} for substance Y is 0.2. For the chromatographic peak, if this ratio is substantially different from the value calculated from the spectra then the peak we are looking at may not be what we think it is, or another substance may be coeluting. The PDA can plot a continuous ratio across the peak which, for a chromatographically pure peak, gives a rectangular shape whose height is proportional to the ratio of the absorptivities at the two wavelengths used. The method is illustrated in Fig. 5.4d.

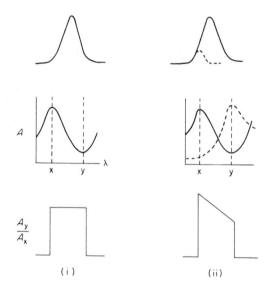

Fig. 5.4d. *Absorbance ratios for (i) a chromatographically pure peak and (ii) a peak with an impurity at the leading edge*

As well as these graphical methods, there are also numerical formats that are used to evaluate data from the PDA. The purity parameter $(\bar{\lambda}_w)$, developed by Varian, reduces the data from a spectrum to a single number by calculating a mean wavelength weighted towards higher absorbance wavelengths (the weighting is done to reduce the effect of noise on the calculation). This data format can be applied to spectra to confirm peak identity by comparison of $\bar{\lambda}_w$ of a peak spectrum with that of a spectrum from a reference compound determined under the same chromatographic conditions. Peak purity can be ascertained by comparing $\bar{\lambda}_w$ values for spectra collected at several points across the peak of interest.

SAQ 5.4a

Fig. 5.4e(i) shows the chromatogram of three anti-convulsant drugs determined in a blood sample. A PDA was used as the detector, with detection wavelength 210 nm. Two of the peaks are partly obscured by excipients (this means all the assorted rubbish in the sample that we are not interested in). Figs. 5.4e(ii), (iii) and (iv) show the UV spectra of each of the drugs. Chromatograms like this can often be improved by changing the detection wavelength as the chromatogram proceeds (i.e. by the use of a wavelength programme). On the basis that the excipients are more likely to absorb UV the shorter the wavelength, suggest a suitable wavelength programme that would provide an acceptable sensitivity for peak 1 and would also discriminate against the interfering peaks.

$\longrightarrow$

SAQ 5.4a
(cont.)

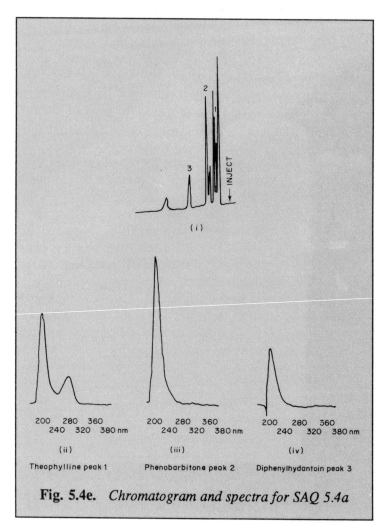

Fig. 5.4e.   *Chromatogram and spectra for SAQ 5.4a*

5.5. FLUORESCENCE DETECTORS

Many compounds are capable of absorbing UV radiation and subsequently emitting radiation of a longer wavelength, either instantly (fluorescence) or after a time delay (phosphorescence). Usually, the fraction of the absorbed energy that is re-emitted is quite low, but for a few compounds values of 0.1–1 are obtained, and such compounds are suitable for fluorescence detection. Compounds that fluoresce naturally have a conjugated cyclic structure e.g. polynuclear (polycyclic) aromatic hydrocarbons. Many non-fluorescent compounds can be converted to fluorescent derivatives by treatment with suitable reagents.

A block diagram of a fluorescence detector is shown in Fig. 5.5a.

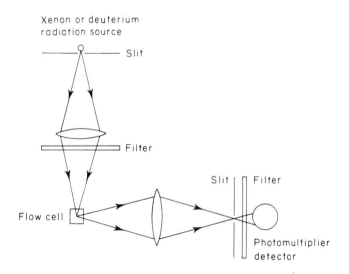

Fig. 5.5a. *Fluorescence detector*

Radiation from a xenon or deuterium source is focussed on the flow cell. An interchangeable filter allows different excitation wavelengths to be used. The fluorescent radiation is emitted by the sample in all directions, but is usually measured at 90° to the incident beam. In some types, to increase sensitivity, the fluorescent radiation is reflected and focused by a parabolic mirror. The second filter isolates a suitable wavelength from the fluorescence spectrum and prevents any scattered light from the source from reaching the photomultiplier detector. The 90° optics allow monitoring of the incident beam as well, so that dual UV absorption and fluorescence detection is possible, and some commercial models have this facility.

Fluorescence detectors can be much more sensitive than UV absorbance detectors; for favourable solutes (such as anthracene) the noise equivalent concentration can be as low as 10^{-12} g cm^{-1}. Because both the excitation wavelength and the detected wavelength can be varied, the detector can be made highly selective, which can be very useful in trace analysis. The response of the detector is linear provided that no more than about 10% of the incident radiation is absorbed by the sample. This results in a linear range of 10^3–10^4.

Polycyclic aromatic hydrocarbons (PAH) are important air pollutants that have to be detected at very low concentrations. Fig. 5.5b shows the separation of a synthetic mixture of very low levels of PAH. They are barely detectable using UV absorption, but are easily monitored by fluorescence.

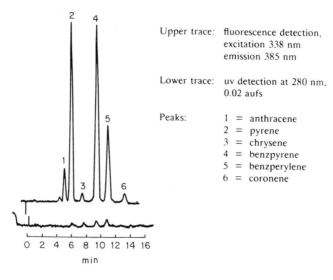

Upper trace: fluorescence detection, excitation 338 nm emission 385 nm

Lower trace: uv detection at 280 nm, 0.02 aufs

Peaks: 1 = anthracene
         2 = pyrene
         3 = chrysene
         4 = benzpyrene
         5 = benzperylene
         6 = coronene

Fig. 5.5b. *Separation of PAH*

5.6. ELECTROCHEMICAL (EC) DETECTORS

5.6.1. Introduction

Electrochemical detectors measure either the conductance of the eluent or the current associated with the oxidation or reduction of solutes. To be capable of detection using the first method the solutes must be ionic, and using the second method the solutes must be relatively easy to oxidize or reduce.

The first type are called conductivity detectors and are used for the detection of inorganic or organic ions, usually after separation by ion exchange chromatography. The use of conductivity detectors is dealt with in Section 7.6.6. Electrochemical detectors that measure current associated with the oxidation or reduction of solutes are called amperometric or coulometric detectors. The term 'EC detector' normally refers to these types rather than to conductivity detectors.

When current is passed through a solution, reactions occur at each electrode in which electron exchange takes place between the electrode and substances in solution. At the cathode, substances gain electrons (reduction) and at the anode they lose electrons (oxidation). We can think of the cathode and anode as a reducing and an oxidizing agent respectively, whose strength depends on the value of the electrode potential. A cathode becomes a stronger reducing agent as its potential becomes more negative and an anode becomes a stronger oxidizing agent as its potential becomes more positive. A substance that can be electrochemically oxidized or reduced is said to be electroactive. If it is difficult to oxidize or reduce chemically, it will be difficult to do so electrochemically as well, so that the reduction will need a cathode with a relatively large negative potential or the oxidation an anode with a relatively large positive potential. Because we are dealing with electroactive substances in solution, we always have to consider the possible electroactivity of the solvent.

In Fig. 5.6a the potential E of an electrode (measured against a suitable reference electrode) is plotted against the current flowing in the cell. Because oxidation and reduction result in different directions of current flow (for reduction electrons flow into the electrode and for oxidation electrons flow out of it) we distinguish between the two by calling a reduction current (a cathodic current) positive and an oxidation (anodic) current negative. The solid line in the figure shows the current–potential curve obtained for a solution containing electroactive solutes A and B which are oxidizable and C which is reducible. The broken line shows the background current (i.e. the current–potential curve of the solvent in the absence of A, B and C).

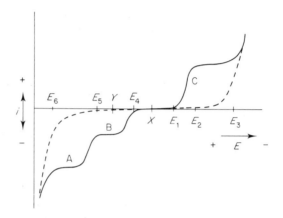

Fig. 5.6a. *Current–potential curves for three electroactive solutes A, B and C*

If we start at point X on the graph and make E increasingly negative by moving to the right, at E_1 we observe a current as C is reduced. After E_2 the current is fairly constant (the limiting current). In this region C is reduced as soon as it reaches the electrode, so that the rate of reduction of C, and therefore the current, is limited by the rate at which C can get to the electrode from other parts of the solution, i.e. by its concentration. At E_3 there is a large increase in current as the solvent is reduced. Similar considerations apply to the anodic part of the graph. Oxidation of B (easier than A) begins at E_4 and of A at E_5 (the anodic current of A is superimposed on the limiting current of B). Finally the solvent is oxidized at E_6.

For EC detection we measure the current in a flow cell at the column outlet. We can change the selectivity of the detector by changing the electrode potential. For instance, if the electrode potential was at X in the figure then none of the compounds would be detected. At Y the electrode would detect B but not A. The electrode potential is being used here to change selectivity in a similar way to a change of wavelength with UV detection.

To use reduction as a method of EC detection is more difficult than oxidation. Oxygen is very easily reduced, and if present in the mobile phase will create a background current much larger than the current due to the solutes. To prevent this, the oxygen has to be carefully removed, which is not easy in practice. Having said this, there is no doubt that as the samples examined by EC methods become progressively more challenging, the use of EC reduction is increasing. Another important consideration with EC detectors is that the mobile phase used must have fairly high conductance, so they are used with aqueous–organic mixtures containing added salts, or with buffer solutions.

∏   For which of the following compounds could EC detection be useful?

Methylbenzene,   decane,   phenol,   nitrobenzene, 2-chloroaniline.

The hydrocarbons are not easily oxidizable or reducible and so would not be suitable. Nitrobenzene would have to be examined by EC reduction— UV detection would probably be easier. Phenols and aromatic amines are easily oxidized and are suitable for EC detection.

5.6.2.   Amperometric Detectors

These detectors oxidize or reduce only a small quantity of the solute (less than 1%), so the currents observed are very small (nanoamps). Such currents are not too difficult to measure using modern amplifiers and the detector has a high sensitivity, considerably higher than that of UV/visible absorbance detectors, although not as good as fluorescence detectors. Noise equivalent concentrations of about 10^{-10} g cm^{-3} have been obtained in favourable cases. Another advantage of these detectors is that they can be made with a very small internal volume.

Fig 5.6b is a simplified diagram of an amperometric detector. Three electrodes are used, called working, auxiliary and reference electrodes (WE, AE and RE). The WE is the electrode at which the electroactivity is monitored, and the RE, usually a silver–silver chloride electrode, provides a stable and reproducible voltage to which the potential of the WE can be referenced. The AE, usually stainless steel, is a current- carrying electrode.

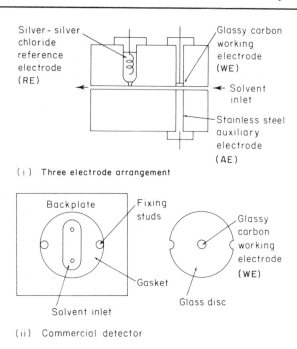

(i) Three electrode arrangement

(ii) Commercial detector

Fig. 5.6b. *Amperometric detector*

The material most commonly used for the WE is glassy carbon, which is a pyrolytically prepared form of carbon that is inert and electrically conducting. The better grades of it have a smooth surface that can accept a high polish. The main problem with solid electrodes like this one is lack of reproducibility caused by degradation of the electrode surface, so that from time to time the electrode surface has to be smoothed, e.g. by lightly polishing with recorder chart paper or cleaning with chromic acid. Because EC flow cells need this regular maintenance, they have to be easy to get into and reassemble. Fig. 5.6b(ii) shows part of the construction of a commercial detector (Waters 460). The glassy carbon WE is embedded in a borosilicate glass disc which is clamped on to a backplate. The cell is formed by a PTFE gasket between the disc and the backplate. The cell volume can be varied by changing the thickness of the gasket (the lowest volume for this one is 2.5 μl).

5.6.3. Coulometric Detectors

The coulometric detector (Coulochem, supplied by Severn Analytical) is a multielectrode device that can use up to four porous graphite working electrodes (each with associated AE and RE). The column eluent flows through these electrodes rather than over them, and with proper choice of potential the detector reacts with all of the electroactive solute passing through it. This is claimed to produce a better signal to noise ratio than is obtained with amperometric detectors, and there is some evidence to suggest that the porous electrodes are less prone to surface degradation than the glassy carbon types.

The working electrodes can be operated at different potentials and the signal from any or all of them can be monitored, so that the electrode arrangement can be used in a number of different modes. For example, with favourable *i–E* curves, interfering or undesirable compounds that are electroactive can be removed at one electrode and compounds of interest detected at another. Fig. 5.6c shows the principle of this, in which a small amount of B is to be determined in a large excess of unresolved component A.

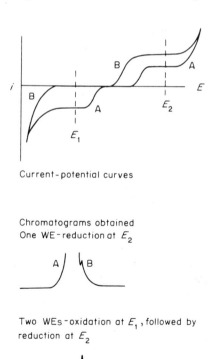

Current-potential curves

Chromatograms obtained
One WE-reduction at E_2

Two WEs-oxidation at E_1, followed by reduction at E_2

Fig. 5.6c. *Coulometric detection using dual electrodes*

Another possibility is to operate two working electrodes at different points on a given *i–E* curve so as to obtain a ratio for the peak in question. This can be used as a criterion of peak purity by comparison with the ratio obtained for a standard.

Although they are more sensitive than UV absorbance detectors, EC detectors are not as easy to use, and have a more limited range of applications. They are chosen for trace analyses where the UV detector does not have high enough sensitivity. Fig. 5.6d shows some examples of compounds for which EC detection has been used.

Compound type	Examples
Phenols, amines	Neurotransmitters, e.g. adrenaline, dopamine; amino acids
Heterocyclic nitrogen compounds	Cocaine, morphine alkaloids, phenothiazines, purines
Sulphur compounds	Penicillins, thioureas, amino acids
Unsaturated alcohols	Vitamin C
Anions	I^-, $S_2O_3^{2-}$, SCN^-

Fig. 5.6d.   *Compounds which can be detected by EC*

SAQ 5.6a

Fig. 5.6e shows the current–potential curves of two electroactive solutes X and Y. In a solution containing both of them:

(*i*) Which would be detected by an EC detector operating at a potential E_2?

(*ii*) Which would be detected at E_3?

$\longrightarrow$

SAQ 5.6a
(cont.)

(iii) Operating the detector at a potential E_4 would not be a good idea, and operation at E_1 would not be very smart either. What would be detected at each of these potentials, and what is wrong with the choice of potential in each case?

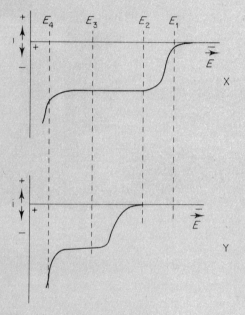

Fig. 5.6e. *Current–potential curves for two electro-active solutes X and Y*

5.7.   REFRACTIVE INDEX (RI) DETECTORS

These detectors sense the difference in refractive index between the column eluent and a reference stream of pure mobile phase. They are the closest thing in HPLC to a universal detector, as any solute can be detected as long as there is a difference in RI between the solute and the mobile phase.

∏   Look at the refractive index values given in Fig. 5.7a and then see if you can answer the questions below.

Hexane	1.375
Octane	1.397
Nonane	1.405
Decane	1.410
Tridecane	1.425
Benzene	1.501
Tetrahydrofuran	1.405

Fig. 5.7a.

(a) If the alkanes in the table were separated using tetrahydrofuran as the mobile phase, for which alkane would an RI detector show the lowest sensitivity?

(b) With tetrahydrofuran as the mobile phase, what would be unusual about the appearance of the chromatogram?

(c) How would the appearance of the chromatogram change if benzene was used as the mobile phase?

(d) With which mobile phase, benzene or tetrahydrofuran, would the RI detector show the greater sensitivity for tridecane?

Refractive index detectors are not as sensitive as UV absorbance detectors. The best noise levels obtainable are about 10^{-7} r.i.u. (refractive index units), which corresponds to a noise equivalent concentration of about 10^{-6} g cm^{-3} for most solutes. The linear range of most RI detectors is about 10^4. If you want to operate them at their highest sensitivity you have to have very good control of the temperature of the instrument and of the composition of the mobile phase. Because of their sensitivity to mobile phase composition it is very difficult to do gradient elution work, and they are generally held to be unsuitable for this purpose.

Several rather different designs of RI detector have been used in HPLC. Fig. 5.7b shows the operating principle of one type, the deflection refractometer. Light from the source S is focused on to the cell, which consists of sample and reference chambers separated by a diagonal sheet of glass. After passing through the cell, the light is diverted by a beam splitter B to two photocells P1 and P2. A change in the RI of the sample stream causes a change in the angle at which the beam strikes the splitter. This causes a change in the relative amounts of light falling on P1 and P2, and therefore a difference in their relative output. This difference is amplified, giving an error signal at the amplifier output that operates a servomotor, which rotates the beam splitter until the error signal is reduced to zero. The beam splitter movement (proportional to the difference in RI that caused it) is measured by the recorder.

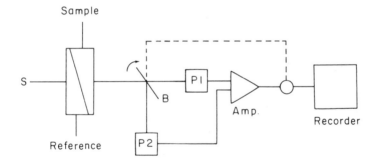

Fig. 5.7b. *Refractive index detector (refractometer)*
A change in the RI of the sample stream alters the output of P1 and P2, producing a signal at the amplifier output that operates a null-balance system

SAQ 5.7a	For which of the following analyses do you think that UV absorbance detection would not be suitable? If UV absorbance is unsuitable, suggest an alternative detector. $\longrightarrow$

SAQ 5.7a
(cont.)

(*i*) The determination of mixed sulphonamide drugs in a tablet.

(*ii*) The separation of polyethene into fractions of different relative molecular mass, using exclusion chromatography.

(*iii*) The determination of phenols as contaminants in a sample of river water.

(*iv*) The analysis of B vitamins in a multivitamin tablet.

(*v*) The determination of riboflavin (vitamin B_2) in milk.

The general structure of the sulphonamides is:

$$R_1 - NH - \langle \text{ring} \rangle - SO_2NH - R_2$$

Structures of some B vitamins

Thiamine (B_1)

$$\left[H_3C \underset{N}{\overset{N}{\diagdown}} \overset{NH_2}{\diagup} \quad \underset{\underset{CH_2^+}{|}}{\overset{S}{\diagdown}} \overset{CH_2CH_2OH}{\diagup} \right] Cl^-$$

Pyridoxine (B_6)

$$H_3C \underset{HO}{\overset{N}{\diagdown}} \overset{CH_2OH}{\diagup} \\ CH_2OH$$

$\longrightarrow$

SAQ 5.7a
(cont.)

Niacinamide

Riboflavin (B$_2$)

5.8. DERIVATIVE PREPARATION

In GC, derivatives are usually prepared to allow or improve the chromatography of the sample. The purpose of derivative preparation in HPLC is usually to improve detection, especially when determining traces of solutes in complex matrices, such as biological fluids or environmental samples.

Derivative preparation can be performed either prior to the separation (pre-column derivatization) or after (post-column derivatization), and can be done either on-line or off-line. The two techniques most commonly used are pre-column off-line and post-column on-line.

The first of these requires no modification to the instrument and, compared with post-column techniques, has fewer limitations as regards reaction time and conditions. On the other hand, the formation of a reasonably stable and well-defined product is necessary, the presence of excess reagent and by-products may interfere with the separation, and derivatization may alter the properties of the sample that facilitated separation.

Post-column on-line derivatization is carried out in a reactor located between the column and the detector. With this technique, the derivatization reaction does not need to go to completion, provided that it can be done reproducibly, and the reaction does not produce any chromatographic inter-ferences. The reaction needs to take place in a fairly short time at moderate temperatures, and the reagents should not be detectable under the same conditions at which the derivative is detected. The mobile phase may not be the best medium in which to carry out the reaction, and the presence of the reactor after the column will increase the extra-column dispersion.

Fig 5.8a shows three types of post-column reactor. In the open tubular reactor, after the solutes have been separated on the column, reagent is pumped into the column effluent via a suitable mixing tee. The reactor, which may be a coil of stainless steel or PTFE tube, provides the desired holdup time for the reaction. Finally, the combined streams are passed through the detector. This type of reactor is commonly used in cases where the derivatization reaction is fairly fast. For slower reactions, segmented stream tubular reactors can be used. With this type, gas bubbles are introduced into the stream at fixed time intervals. The object of this is to reduce axial diffusion of solute zones, and thus to reduce extra-column

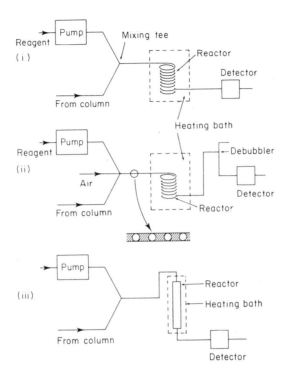

Fig. 5.8a. *Post-column reactors*

(i) Open tubular reactor
(ii) Segmented reactor
(iii) Packed bed reactor

dispersion. For intermediate reactions, packed bed reactors have been used, in which the reactor may be a column packed with small glass beads.

The reagents used in post-column reactors are:

(*a*) Fluorotags: non-fluorescent molecules that react with solutes to form fluorescent derivatives.

(*b*) Chromatags: which form a derivative that strongly absorbs UV or visible radiation.

Example of a few of the more popular reagents are given in Fig. 5.8b.

Structure	Used for	Conditions for detection of derivative
Fluorotags		
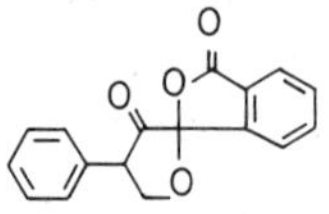	Compounds containing primary nitrogen, e.g. amines, amino acids, peptides	Excitation 390 nm Emission 470 nm
Fluorescamine		
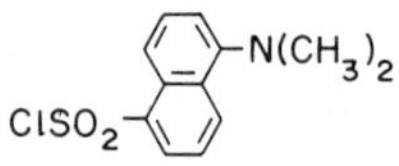	Proteins, amines, amino acids, phenolic compounds	Excitation 335–365 nm Emission 520 nm
Dansyl chloride		
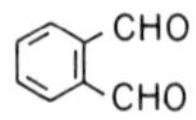	Compounds containing primary nitrogen	Excitation 300 nm Emission 400–600 nm
OPA		
Chromatags		
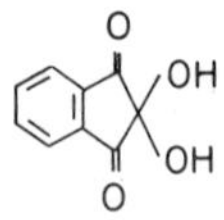	Amino acids	Absorption at about 570 nm
Ninhydrin		
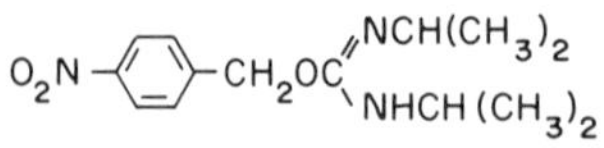 PNBDl	Carboxylic acids	Absorption at about 254 nm

Fig. 5.8b.   *Reagents used in post-column reactors*

It is sometimes possible to improve detection by changing the pH of the eluent, or by the use of photochemical reactions. The common barbiturates used in therapy are weak acids that are easily separated in their acid (non-ionized) forms. Because the conjugate bases are much stronger chromophores than the acids, barbiturates have been detected by post-column mixing with a pH 10 borate buffer followed by UV absorption measurement at 254 nm. An example of the second approach is the detection of cannabis derivatives in body fluids, involving the conversion of cannabis alcohols to fluorescent derivatives on subjecting the column effluent to intense UV radiation.

SAQ 5.8a	Consider a solute which is detected by derivatization using a post-column reactor of the type shown in Fig. 5.8a.

What would be the effect on the peak area of this solute of:

(*i*) Increasing the length of the reactor coil?

(*ii*) Increasing the temperature of the reactor coil?

(*iii*) Increasing the flow rate of reagents into the mixing tee?

What would be the effect on the resolution between two peaks in the chromatogram of increasing the length of the reactor coil?

SAQ 5.8a

5.9. SUMMARY

A large number of devices have been used as detectors for HPLC. The characteristics of five of the more important types are described in Chapter 5, and examples are given of the range of samples for which they can be used. The use of derivative preparation as an aid to detection is considered.

Learning Objectives

Now that you have completed Chapter 5 you should be able to:

- Specify the properties that are required of an HPLC detector.

- Understand the operating principles and the limitations of the important types of detector.

- Recognize the samples for which different detectors can be used.

- Describe how derivatives can be prepared to improve detection.

References

GENERAL REVIEWS OF DETECTORS

1. R.P.W. Scott, *Liquid Chromatography Detectors*, Elsevier, 1986.

2. E.S. Yeung, Ed. *Detectors for Liquid Chromatography*, Wiley, 1986.

3. E.S. Yeung and R.E. Synovec, *Analytical Chemistry*, 1986, 58, 1237A–1256A.

4. C.F. Simpson, Ed. *Techniques in Liquid Chromatography*, Wiley, 1984, Chapter 6.

5. J.W. Dolan, *LC–GC*, 1986, 4, 526–529.

PHOTODIODE ARRAY

6. T. Alfredson and T. Sheehan, *Journal of Chromatographic Science*, 1986, 24, 473–482.

7. A.J. Owen, *The Diode Array Advantage in UV/Visible Spectroscopy*, Anachem Ltd., Charles St., Luton LU2 0EB.

8. M.V. Pickering, *LC–GC International*, 1991, 4(1), 20–25.

ELECTROCHEMICAL

9. P.T. Kissinger, *Analytical Chemistry* 1977, 49, 447A–456A.

DERIVATIVE PREPARATION

10. R.W. Frei, H. Jansen and U.A.T. Brinkman, *Analytical Chemistry* 1985, 57, 1529A–1539A.

SAQS AND RESPONSES

SAQ 5.2a

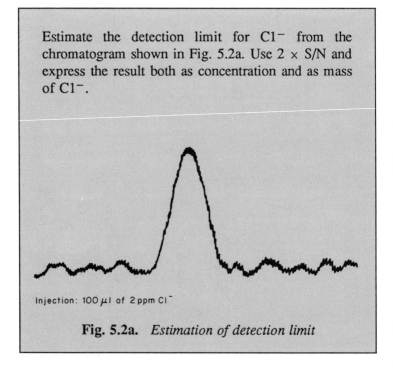

Estimate the detection limit for Cl^- from the chromatogram shown in Fig. 5.2a. Use 2 × S/N and express the result both as concentration and as mass of Cl^-.

Injection: 100 μl of 2 ppm Cl^-

Fig. 5.2a. *Estimation of detection limit*

Response

The measurements to be made are shown in Fig. 5.2b. The baseline taken is an average value, drawn through the noise.

Baseline noise = 5 mm
Peak height   = 40 mm

An injection of 100 μl of 0.5 ppm would give a peak height of 10 mm (twice baseline noise). This injection would contain 50 ng of Cl^-.

The detection limit is quoted as:

0.5 ppm Cl^- for 100 μl injection, using 2 × S/N, or 50 ng Cl^-, using 2 × S/N.

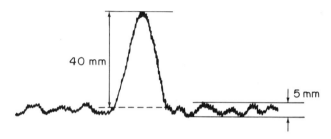

40 mm

5 mm

Injection : 100 μl of 2 ppm Cl^-

Fig. 5.2b. *Estimation of detection limit*

****

SAQ 5.3a

Fig. 5.3c shows the UV spectra of azobenzene (Az, concentration 3.73×10^{-3} g dm^{-3}) and phenanthrene (P, 3.23×10^{-3} g dm^{-3}) both recorded in *iso*-octane on a standard UV/visible spectrophotometer. The wavelength drive on the instrument was 10 nm cm^{-1} and the absorbance range was 2 a.u.f.s. Measurements were made against *iso*-octane using 10 mm cells.

What wavelength would you choose:

  (*i*) To detect Az without detecting P;

  (*ii*) To detect P without detecting Az;

  (*iii*) To detect both of them;

  (*iv*) To detect Az at maximum sensitivity? $\longrightarrow$

SAQ 5.3a
(cont.)

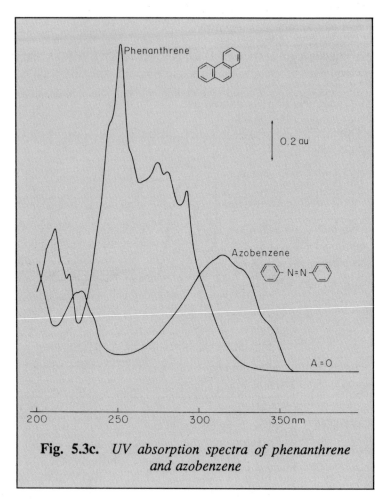

Fig. 5.3c. *UV absorption spectra of phenanthrene and azobenzene*

Response

(*i*) The shoulder at 342 nm.

(*ii*) It cannot be done, but at 251 nm the ratio of P to Az sensitivity would be greatest.

(*iii*) Between 270 and 320 nm.

(*iv*) Az has a maximum absorbance at 314 nm at this wavelength.

****

SAQ 5.3b

Infrared absorption detectors are available for HPLC, although they have never become very popular. From what you know about IR spectrometry and what you have read so far about HPLC detectors, see if you can decide whether the following statements are true or false.

(*i*) An infrared spectrum provides more structural information about a compound than does a UV spectrum;

(*ii*) An IR detector would be more sensitive than a UV detector;

(*iii*) An IR detector could not be used with solvent mixtures containing water;

(*iv*) An IR detector could be used as a selective detector or a universal detector by changing the wavelength used.

Response

   (*i*) True.

   (*ii*) False. If you have ever done any practical IR work you should have had no trouble with these two. An infrared spectrum provides a great deal of structural detail, but the method does not have very high sensitivity.

   (*iii*) False. The flow cell would have to be made from a water insoluble material that is transparent to infrared radiation, e.g. KRS5 (TlBr/TlI). Glass cannot be used for optical components in IR instruments, as it absorbs IR radiation.

   (*iv*) True with qualifications. The detector could be used selectively by operating at, for instance, 1725 cm^{-1}, where it would detect some solutes containing carbonyl groups, or as a universal detector at somewhere in the CH stretching region. The problem with universal detection would be to find a wavelength that was not absorbed by the mobile phase. The detector would have to be operated in 'windows' in the IR spectrum of the mobile phase; the number of suitable solvents is very limited.

****

SAQ 5.4a	Fig. 5.4e(i) shows the chromatogram of three anti-convulsant drugs determined in a blood sample. A PDA was used as the detector, with detection wavelength 210 nm. Two of the peaks are partly obscured by excipients (this means all the assorted rubbish in the sample that we are not interested in). Figs. 5.4e(ii), (iii) and (iv) show the UV spectra of each of the drugs. Chromatograms like this can often be improved by changing the detection wavelength as the chromatogram proceeds (i.e. by the use of a wavelength programme). On the basis that the excipients are more likely to absorb UV the shorter the wavelength, suggest a suitable wavelength programme that would provide an acceptable sensitivity for peak 1 and would also discriminate against the interfering peaks.

SAQ 5.4a
(cont.)

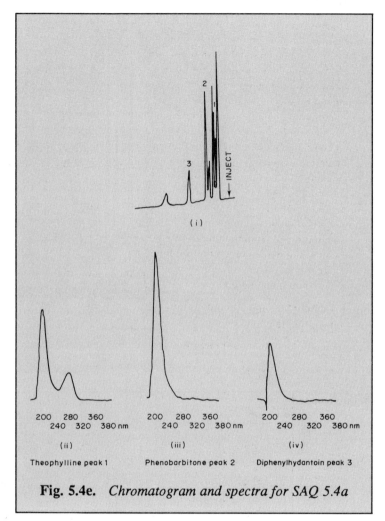

Fig. 5.4e. *Chromatogram and spectra for SAQ 5.4a*

Response

280 nm could be used until just before peak 2 elutes, then the wavelength could be changed to 210 nm. The result of doing this is shown in Fig. 5.4f.

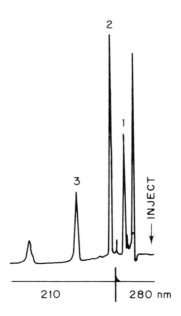

Fig. 5.4f. *Chromatogram of anticonvulsant drugs using wavelength programme*

****

SAQ 5.6a

Fig. 5.6e shows the current–potential curves of two electroactive solutes X and Y. In a solution containing both of them:

(*i*) Which would be detected by an EC detector operating at a potential E_2?

(*ii*) Which would be detected at E_3?

(iii) Operating the detector at a potential E_4 would not be a good idea, and operation at E_1 would not be very smart either. What would be detected at each of these potentials, and what is wrong with the choice of potential in each case?

$\longrightarrow$

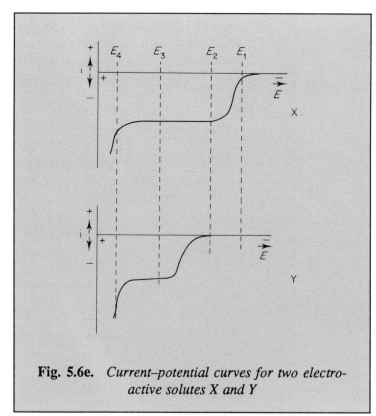

Fig. 5.6e. *Current–potential curves for two electro-active solutes X and Y*

Response

Operation at E_2 would detect X, operation at E_3 would detect both X and Y. Operation at E_4 would detect both X and Y, but at this point the solvent or background electrolyte is oxidized as well. At best there would be a large background current, and it might be impossible to get the recorder on scale. At E_1, X would be detected but the sensitivity would be low. It would be much better to work at a potential on the limiting current plateau, such as E_2.

****

SAQ 5.7a

For which of the following analyses do you think that UV absorbance detection would not be suitable? If UV absorbance is unsuitable, suggest an alternative detector.

(*i*) The determination of mixed sulphonamide drugs in a tablet.

(*ii*) The separation of polyethene into fractions of different relative molecular mass, using exclusion chromatography.

(*iii*) The determination of phenols as contaminants in a sample of river water.

(*iv*) The analysis of B vitamins in a multivitamin tablet.

(*v*) The determination of riboflavin (vitamin B_2) in milk.

The general structure of the sulphonamides is:

$$R_1 - NH -\!\!\left\langle\!\!\bigcirc\!\!\right\rangle\!\!- SO_2NH - R_2$$

Structures of some B vitamins

Thiamine (B_1)

$$\left[\begin{array}{c} H_3C \underset{N}{\overset{N}{\diagdown}} \overset{NH_2}{\diagup} \quad S \diagdown CH_2CH_2OH \\ \\ N \diagdown \underset{CH_2^+}{\overset{N}{\diagup}} CH_3 \end{array} \right] Cl^-$$

$\longrightarrow$

SAQ 5.7a
(cont.)

Pyridoxine (B$_6$)

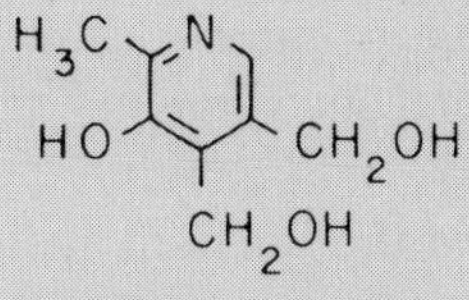

Niacinamide

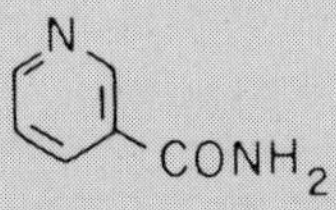

Riboflavin (B$_2$)

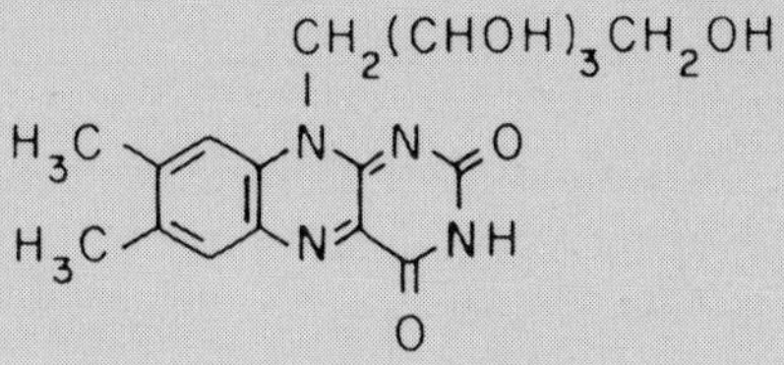

Response

(*i*) These are aromatic, and so will absorb UV radiation. In a tablet there should be reasonably high levels of the compounds present, so UV absorbance detection would be the method of choice.

(*ii*) These compounds are saturated and will not show any UV absorption above 200 nm. A refractometer is the only suitable detector.

LIVERPOOL JOHN MOORES UNIVERSITY
LEARNING SERVICES

(*iii*) The amounts of phenols present are likely to be very low. Trace phenols in water have been determined using both UV absorbance and EC detectors. The sensitivity of the UV absorbance detector is not really high enough, so that sample preconcentration methods have to be used. With the more sensitive EC detection, the analysis can be done without preconcentration.

(*iv*) The structure of the vitamins indicates that they will all absorb UV radiation, so that for reasonably high levels of them in a tablet UV absorbance detection would be suitable.

(*v*) For riboflavin in milk, the low level of vitamin present might be a problem using a UV absorbance detector. Riboflavin has a highly conjugated structure, and will fluoresce, so that fluorescence detection could be used for trace amounts of this compound. The ring nitrogens in the structure indicate that EC detection would be possible as well.

****

SAQ 5.8a

Consider a solute which is detected by derivatization using a post-column reactor of the type shown in Fig. 5.8a.

What would be the effect on the peak area of this solute of:

(*i*) Increasing the length of the reactor coil?

(*ii*) Increasing the temperature of the reactor coil?

(*iii*) Increasing the flow rate of reagents into the mixing tee?

What would be the effect on the resolution between two peaks in the chromatogram of increasing the length of the reactor coil?

Response

(*i*) Unless the derivatization reaction is very fast, an increase in the length of the reactor coil should increase the peak area, because the longer the time the solute spends in the reactor, the more product should be formed.

(*ii*) An increase in the temperature of the reactor should increase the reaction rate and thus increase the peak area.

(*iii*) Changing the flow rate of the reagent may affect the peak area for several reasons. At very low flow rates there may be insufficient reagent to complete the reaction, in which case an increase in flow rate will increase the peak area. But as we increase the flow rate we are diluting our solute and also reducing the time the solute spends in the reactor, and these factors may reduce the response at high flow rates.

An increase in the length of the reactor tube will increase the dispersion and so decrease the resolution between a given pair of solutes. In practice, the length of the reactor coil used will represent a compromise between detector response and resolution.

****

6.   The Mobile Phase

6.1.   THE IMPORTANCE OF POLARITY IN HPLC

The relative distribution of a solute between two phases is determined by
the interactions of the solute species with each phase. The relative strengths
of these interactions are determined by the variety and the strengths of the
intermolecular forces that are present, or, in more general terms, by the
polarity of the sample and that of the mobile and stationary phases.

Intermolecular forces may be caused by a solute molecule having a dipole
moment, whereby it can interact selectively with other dipoles, or if a
molecule is a good proton donor or acceptor it can interact with other
such molecules by hydrogen bonding. Molecules can also interact via much
weaker dispersion forces, which rely on a given molecule being polarized
by another molecule.

Polarity is a term that is used in chromatography as an index of the ability
of compounds to interact with one another in these various ways. It is
applied very freely to solutes, stationary phases and mobile phases. The
more polar a molecule, the more strongly it can interact with other molecules
through the mechanisms above. If the polarities of stationary and mobile
phases are similar then it is likely that the interactions of solutes with
each phase may also be similar, leading to poor separations. Thus for
hydrocarbon-type (non-polar) stationary phases we need a polar mobile
phase, whereas unmodified silica, which is highly polar, needs a mobile
phase with relatively low polarity. If we are concerned with the separation of
solutes that are chemically very similar, we should try to choose a stationary
phase that is chemically similar to our solutes. The retention of solutes is
usually altered by changing mobile phase polarity.

6.2. MEASURING POLARITY

It is easy to see that, for instance, water is a more polar solvent than heptane. Water has a dipole moment, is both a proton donor and acceptor, and will dissolve ionic solutes. Similarly, methanol and acetonitrile are both more polar than heptane, but it is not so easy to assign relative polarities to methanol and acetonitrile.

It is helpful in LC to have a quantitative measure of polarity so that, for example, the relative polarity of a solvent or of a mixture of solvents can be expressed as a number. This may be done in a number of ways, none of which are entirely satisfactory, but they do allow us to arrange solvents in order of polarity and to estimate the polarity of solvent mixtures. One such way is to use as a measure of polarity a quantity called the solubility parameter (δ), defined by:

$$\delta = \left(\frac{\Delta E}{V}\right)^{1/2} \qquad (6.2a)$$

where ΔE = internal energy of vaporization and V = molar volume.

In SI units, δ is measured in $J^{1/2}m^{-3/2}$ or $Pa^{1/2}$, although many authors still give δ in the non-SI $cal^{1/2}$ $cm^{-3/2}$ (the conversion is 1 $cal^{1/2}cm^{-3/2}$ = 2.044 $\times$ 10^3 $Pa^{1/2}$). Another measure of polarity (the polarity index, P') is based on experimental gas chromatographic distribution coefficients for three test solutes on a large number of stationary phases (the method of calculation is beyond the scope of this book). A third method, known as solvatochromic polarity measurement, measures spectral band shifts in the UV/visible region, which, for certain compounds, vary strongly with polarity.

Fig. 6.2a shows δ and P' values for a number of solvents, arranged in order of increasing δ.

∏ Suggest an explanation for the high solubility parameter of water

Water has both strong intermolecular forces, due to hydrogen bonding, and therefore a relatively large ΔE together with a relatively low molar volume.

You can see from the figure that the order of polarity is not quite the same by the two methods, for example the polarity index describes acetonitrile as

Solvent	δ, Pa$^{1/2} \times 10^{-3}$	P'
hexane	14.9	0.1
methylbenzene	18.2	2.4
tetrahydrofuran	18.6	4.0
trichloromethane	19.0	4.1
butanone	19.0	4.7
ethyl ethanoate	19.6	4.4
dichloromethane	19.8	3.1
1,2-dichloroethane	20.0	3.5
propanone	20.2	5.1
1,4-dioxane	20.4	4.8
2-methoxyethanol	23.3	5.5
acetonitrile	23.9	5.8
ethanol	25.9	4.3
methanol	29.4	5.1
water	47.8	10.2

Fig. 6.2a. *Solubility parameter and polarity index for a range of solvents*

more polar than methanol, whereas the solubility parameter (and practical experience) suggests the reverse.

On the basis of solubility parameter alone, solvents with similar values of δ might be expected to show similar solubility properties.

∏   Select some examples from the figure to show that this is not necessarily true.

You could choose methylbenzene ($\delta = 18.2$, immiscible with water) and tetrahydrofuran ($\delta = 18.6$, miscible with water in all proportions) or dioxane ($\delta = 20.4$), propanone ($\delta = 20.2$), both water miscible and dichloroethane ($\delta = 20$, water-immiscible). Differences like these between solvents having the same polarity are called specific effects; the reason for them is that δ measures the total intermolecular interactions that contribute to ΔE. For two solvents, the same total may be composed of different individual contributions; for example, the intermolecular forces in one solvent might be due mainly to hydrogen bonding and in another mainly to dispersion forces. To improve the model we need to be able to specify not only the total intermolecular interactions, but the individual contributions to the total as well. One useful approach to this is the Snyder classification, discussed briefly below.

6.3. THE SNYDER CLASSIFICATION

This scheme classifies solvents on the basis of P' values, and also takes into account the possibility of specific effects. Each solvent is assigned three classification parameters: x_e (proton acceptor parameter), x_d (proton donor parameter) and x_n (strong dipole parameter). The classification then separates solvents into eight groups based on similarity of the x-parameters. Solvents with similar overall polarity (as measured by P' or δ) will now be expected to show similar behaviour only if they are in the same group. Fig. 6.3a shows a selection from each group (solvents with very low polarity are not assigned parameters).

The classification was obtained by studying 81 solvents, and the results are conveniently represented on a solvent selectivity triangle, as shown in Fig. 6.3b.

The eight partially overlapping circles represent the x-parameter regions occupied by the different classes. The method of positioning a solvent (acetonitrile) on the diagram is also shown. The strength of the method is that it enables us to group a large number of chemically different solvents into a limited number of selectivity classes. One practical consequence of this is that, if a certain solvent does not provide sufficient selectivity in a given separation, it is unlikely that another solvent in the same group will do so either.

If we wish to choose a mobile phase consisting of three solvents, then we want to choose solvents having selectivity differences that are as large as possible, so that we can fully exploit these differences by varying the mobile phase composition. Thus we would choose our solvents from groups as far away from each other as possible on the triangle (remember that there are always other factors to consider as well, such as solubility, viscosity, UV cutoff and toxicity). For a reverse phase separation we might choose methanol, acetonitrile and tetrahydrofuran, with a fourth solvent

Solvent	Group	P'	x_e	x_d	x_n	Chemical type
hexane	I	0.1				aliphatic hydrocarbons
diethyl ether	I	2.8	0.53	0.13	0.34	ethers
methanol	II	5.1	0.48	0.22	0.31	aliphatic
ethanol	II	4.3	0.52	0.19	0.29	alcohols
tetrahydrofuran	III	4.0	0.38	0.20	0.42	pyridines, THF
dimethyl sulphoxide	III	7.2	0.39	0.23	0.39	sulphoxides
ethanoic acid	IV	6.0	0.39	0.31	0.30	aromatic alcohols
benzyl alcohol	IV	5.7	0.40	0.30	0.30	acids
dichloromethane	V	3.1	0.29	0.18	0.53	aliphatic halogen
1,2-dicholorethane	V	3.5	0.30	0.21	0.49	compounds
1,4-dioxane	VI	4.8	0.36	0.24	0.40	aliphatic ketones and esters
acetonitrile	VI	5.8	0.31	0.27	0.42	dioxane, nitrites
ethyl ethanoate	VI	4.4	0.34	0.23	0.43	sulphones
propanone	VI	5.1	0.35	0.23	0.42	
methylbenzene	VII	2.4	0.25	0.28	0.47	aromatic hydrocarbons, nitro-compounds
chlorobenzene	VII	2.7	0.23	0.33	0.44	aromatic ethers,
nitroethane	VII	5.2	0.28	0.29	0.43	halogen-substitute aromatic hydrocarbons
water	VIII	10.2	0.37	0.37	0.25	water, fluoroalkanols

Fig. 6.3a. *Selectivity parameters for a range of solvents*

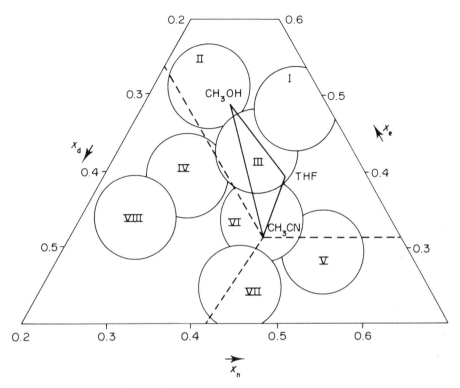

Fig. 6.3b. *Solvent selectivity triangle*

(water) added to adjust the polarity so as to get the k' values of our solutes in the required range. The proportions of these four solvents that produce the optimum separation are then determined by experiment (how this can be done is outlined later in the chapter). In practice, many HPLC separations can be carried out using simple binary mixtures as mobile phases (e.g. methanol/water or acetonitrile/water for reverse phase separations), but as more complex samples are analysed by HPLC the use of ternary and quaternary solvent mixtures as mobile phases will increase. Another strong incentive is that gradient runs can often be replaced by an isocratic separation using an optimized ternary or quaternary mixture.

6.4. ISOELUOTROPIC MOBILE PHASES

The solubility parameter of a mixture of solvents (δ_m) can be estimated using the simple rule:

$$\delta_m = \sum_i \Phi_i \delta_i \qquad (6.4a)$$

where Φ_i = volume fraction and δ_i = solubility parameter of component i.

Thus we can make a mixture of a given polarity using one or more solvents with a higher polarity with one or more solvents having a lower polarity than that required. For example, for a binary mixture of methanol (M) and water (W):

$$\delta_m = \Phi_W \delta_W + \Phi_M \delta_M$$

$$= \delta_W(1 - \Phi_M) + \Phi_M \delta_M \quad \text{(since } \Phi_M + \Phi_W = 1\text{)}$$

$$= \delta_W - \Phi_M(\delta_M - \delta_W) \qquad (6.4b)$$

In reverse phase chromatography, mixtures of different solvents having the same overall polarity are called isoeluotropic mixtures (i.e. having the same eluting strength). In practice it is found that the replacement of a mobile phase by an isoeluotropic mixture will produce roughly the same analysis time, but the k' values of individual components of the mixture will vary, owing to the specific effects mentioned earlier, thus altering the resolution.

SAQ 6.4a	Calculate the composition of binary tetrahydrofuran/water and acetonitrile/water mixtures that are isoeluotropic with methanol/water 50:50.

SAQ 6.4a

A ternary solvent mixture can be represented as a point on a triangular diagram, and a quaternary mixture as a point within a tetrahedron, as shown in Figure 6.4a. In the tetrahedral diagram each vertex represents a single solvent, edges are binary mixtures and each face is one of the four possible ternaries. The ternary diagram shows a line that connects isoeluotropic M/W and T/W compositions. Points on this or similar lines represent ternary compositions that are isoeluotropic with the two binaries. Similarly, in the quaternary diagram we can define an isoeluotropic plane with the vertices corresponding to M/W, T/W and A/W isoeluotropic compositions. Any point on this plane will be a quaternary composition isoeluotropic with the binaries.

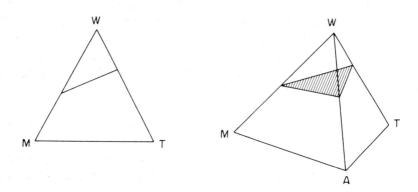

W = water   M = methanol   T = tetrahydrofuran   A = acetonitrile

Fig. 6.4a. *Ternary and quaternary solvent mixtures, showing isoeluotropic regions*

6.5. MOBILE PHASE OPTIMIZATION

When an HPLC method is developed, we usually want to achieve an acceptable degree of separation for all the components of interest in our sample in a reasonable time. We may want to separate all the components with some defined minimum resolution, or to achieve the maximum resolution in a certain time, or we may only be interested in resolving one or two components from the others, and so on. To some extent, the separation can be influenced by operating parameters such as column, temperature and flow rate, but, provided we have made sensible choices for these, by far the most important factor that controls the separation is the composition of the mobile phase.

If a satisfactory separation cannot be achieved using a binary solvent mixture as mobile phase, a ternary or quaternary composition may work. Gradient separations can often be replaced by an isocratic ternary or quaternary composition, as mentioned earlier. If such a suitable composition exists, it will be located somewhere within the triangular or tetrahedral shapes of Fig 6.4a (this is known as the factor space). The problem in mobile phase optimization is finding out where this composition is. Clearly, there is no point in searching the whole of the factor space for the optimum mobile phase composition; for instance, in the tetrahedron of Fig. 6.4a for a reverse phase separation, compositions in the region of the water vertex would lead to unacceptably long analysis times, whereas compositions in the regions of the other three vertices would produce very short analysis times, with inadequate resolution. The first step in an optimization process is usually the reduction of the factor space to, for example, an isoeluotropic line or plane, which is then searched for the optimum.

Mobile phase optimization has been investigated by a number of research groups for about the last ten years; mostly the work has been concerned with reverse phase chromatography. As a result of this work, a number of strategies for optimization have been developed, some of which are discussed briefly below.

6.5.1. Sequential Methods

Most of these are based on the sequential simplex algorithm, which is a multidimensional sequential search procedure used for locating an experimental optimum. A minimum number of initial experiments is performed and, based on these, the algorithm directs the adjustment of experimental conditions away from those which give a poor result and towards those giving a better result. The simplex itself is a geometric shape with one more vertex than the number of variables being studied, thus for two variables it is a triangle, each vertex describing a set of operating conditions. Chromatograms are run at each of these three sets of conditions and the quality of the separation is assessed for each set. The set giving the worst separation is rejected and a new vertex is generated by reflecting the figure through the plane joining the remaining vertices. Repeated application of the procedure leads to a set of operating conditions that are in the region of the optimum set. The procedure is illustrated in Fig. 6.5a.

Initial experiments are carried out using the conditions described by A, B and C. Set B gives the worst result, so it is rejected and the triangle reflects towards point D. Another experiment is done using conditions at D. Out of the three sets A, C and D, A is now rejected and the triangle reflects again, giving another set at E and so on. The simplex proceeds in the direction of experimental conditions that produce more favourable results, and eventually approaches the optimum at O.

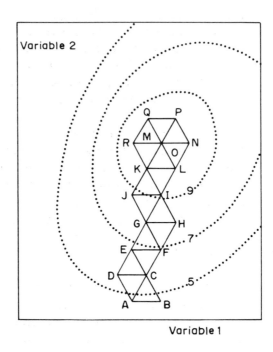

Fig. 6.5a. *Two-variable simplex optimization. The dotted lines are contours representing equal separation quality*

Around the optimum reflection of the triangle will sometimes result in a position at which a measurement has already been made, for example MLN, with N giving the worst result. Instead of rejecting N and returning to measure K again, the second worst result (L) is rejected and the triangle reflects towards P. This procedure is repeated until R is reached, after which no new measurements are suggested by rejecting either the worst or the second worst result, and the process stops after 17 experiments. A two-variable procedure like this has been used for the optimization of ternary solvent mixtures, the two variables being two of the three volume fractions in the mixture. This is sufficient to define the mixture, as the sum of the three volume fractions must equal 1.

An important condition of this and other optimization methods is that an objective criterion of separation quality has to be used, in other words we have to be able to describe the quality of a complete chromatogram by a number. Many different approaches have been used for this. Some of the simpler ones are the resolution between the worst resolved pair of peaks or the sum or product of resolutions between adjacent pairs of peaks. Simplex optimizations tend to use the rather more complex chromatographic response functions, where the quality criterion contains the sum of resolutions between adjacent pairs, but also terms involving the number of peaks detected, the analysis time and several weighting factors.

The two main disadvantages of simplex methods are, first, that a large number of experiments may be required and, second, that the factor space may contain a number of regions where good results are obtained (local optima), which may be homed in on rather than the region where the best result (the global optimum) is found. The second problem can sometimes be resolved by starting the simplex from different initial experimental conditions, but this further increases the number of experiments.

6.5.2. Predictive Methods

These usually begin with a reduction of the factor space, e.g. for a four solvent optimization the tetrahedral factor space is first reduced to an isoeluotropic plane. The most convenient way to do this is to perform an initial gradient run on the sample using 0–100% methanol/water. From this it is possible to calculate a suitable isocratic methanol/water composition that will elute all solutes with k' values in the range 1–10. Once this is known, the two other isoeluotropic binary compositions, e.g. THF/water and acetonitrile/water, are calculated, thus locating the three corners of the isoeluotropic plane. This calculation is done using empirical transfer rules based on the experimental determination of solute retentions in a variety of ternary solvent mixtures. These transfer rules give results slightly different to those obtained using the mixture rule for solubility parameters (Eq. 6.4a and 6.4b).

The isoeluotropic plane represents a region of solvent composition where, in theory, at every point the same analysis time will be produced, although the retention of individual components will vary at different points. Once the position of the plane has been established, further runs are carried out at different points on the plane; a common procedure is to use what is called the seven point statistical design shown in Fig. 6.5b. From the results of these runs, a computer algorithm is used to predict the retention behaviour of each component over the entire surface of the plane. A number of mathematical functions are available for this, each one requiring different numbers of initial runs on the plane. The result of this modelling process is a retention surface (a map of retention behaviour for each component over the surface of the plane). From the retention surface the response surface can be calculated, this being a similar map of chromatographic quality. The optimum mobile phase composition can now be chosen from the response surface.

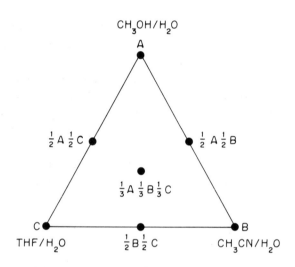

Fig. 6.5b. *Seven point statistical design*

The modelling process used in these methods may suffer from inaccuracy, but the main problem with predictive methods is one of peak recognition. For the experimental results that are used to model the retention map, the retention time of each peak in each chromatogram has to be known. As the mobile phase composition is changed the elution order of peaks may change, or one or more peaks may coelute, or both. This leads to problems in recognizing which peak is which in the various chromatograms. Clearly, this could be established by running standards at each of the mobile phase compositions used, but this is an extremely tedious process. Predictive methods therefore attempt to build peak tracking into the computer software by recognizing (not identifying) peaks from their UV spectra or peak area or a combination of both.

The following is a brief account of an in-house optimization scheme used by ICI Colours and Fine Chemicals Research Centre. The sample is first run in a suitable isocratic methanol/water (or buffer) composition, established by a prior gradient run, and then in the isoeluotropic TFH/water and acetonitrile/water binaries. It may be that the desired separation is achieved in one of these binaries, in which case no further development is needed. If the desired separation has not been achieved and there are only small changes in selectivity in the three binaries, then further mobile phase optimization is probably not justified, and the previous steps are repeated using a different stationary phase (e.g. C-8 instead of C-18). However, if there are marked selectivity differences in the binaries, then optimization is continued using the seven point experimental design shown in Fig. 6.5b, combined with suitable predictive computer software. Fig 6.5c shows results for a sample of anthraquinone and various substituted anthraquinones. Note that the separation is unsatisfactory in each of the isoeluotropic binaries, (i)–(iii), but that large selectivity changes occur. Fig. 6.5c(iv) shows the separation obtained in the predicted optimum quaternary composition.

Optimization packages based on sequential and predictive methods are available commercially, and are discussed in Section 6.5.4.

6.5.3. Iterative Methods

In these methods a simple model is fitted to a few experimental points to predict an optimum composition, which is tested. If the optimum is unsatisfactory, the results of the test are used to improve the model so as to produce a better optimum. In the example below a ternary mobile phase of methanol, THF and water is optimized for the separation of a mixture of six aromatic solutes. A suitable isocratic methanol/water composition producing k' values in the range 1–10 is first established via a methanol/water gradient, as in the previous section. This was calculated as 50% methanol. The sample is now run in this and in an isoeluotropic THF/water composition [Fig 6.5d, chromatograms (i) and (ii)]. This produces two chromatograms with roughly the same analysis time but different k' values for some of the individual peaks. The retention of each component at ternary isoeluotropic compositions is now modelled by assuming a linear variation of $\ln k'$ with volume fraction. Using this assumption, a phase selection diagram is constructed as in (iii), in which a line is drawn connecting the experimental $\ln k'$ values for each solute observed in the two isoeluotropic binaries. These retention data are now used to calculate a criterion of chromatographic quality over the range of mobile phase composition. The criterion used

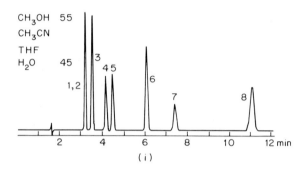

(i)

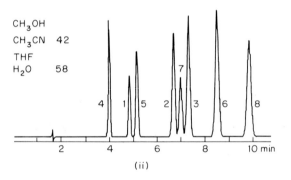

(ii)

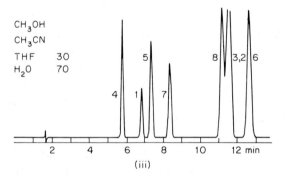

(iii)

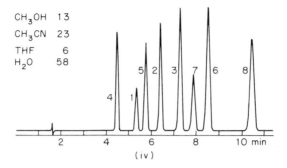

CH$_3$OH 13
CH$_3$CN 23
THF   6
H$_2$O   58

(iv)

1   1,5-diamino
2   1,8-dinitro
3   1,5-dinitro
4   2-amino
5   1,8-diamino
6   1-nitro
7   1-amino
8   anthraquinone

Fig. 6.5c. *Optimization of mobile phase composition for substituted anthraquinones*

here (πR_S) is the product of all resolution values between adjacent peaks in the chromatogram. The variation of πR_S with composition is shown on the phase selection diagram. Any peaks that coelute will give $\pi R_S = 0$, corresponding to any point of intersection of the interpolated $l_n k'$ values. For example, the diagram predicts that a mixture containing 10% THF, 35% methanol and 55% water will produce coelution of peaks 4, 5 and 6. This is shown in chromatogram (iv). The optimum composition, corresponding to maximum πR_S is shown in chromatogram (v).

In this instance the first predicted optimum is satisfactory. Fig. 6.5e shows an example in which the first predicted optimum from the phase selection diagram (i) produces a chromatogram containing unresolved peaks, (ii). The retention data from this chromatogram are now used to produce an improved phase selection diagram, (iii), that predicts a different optimum mobile phase composition, the chromatogram for which is shown in (iv). Both of these examples were taken from ref. 7 at the end of the chapter.

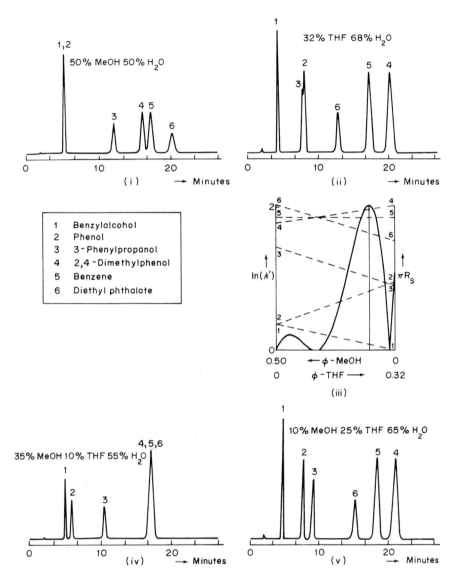

Fig. 6.5d. _Optimization of a ternary mobile phase_

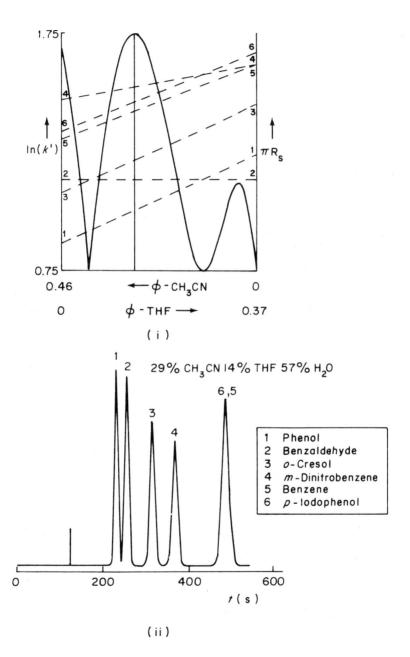

Fig. 6.5e. *Optimization of a ternary mobile phase*

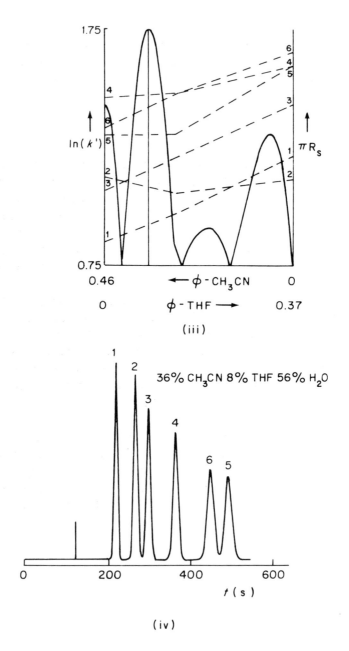

(iii)

(iv)

Fig. 6.5e. (contd) *Optimization of a ternary mobile phase*

6.5.4. Commercial systems

The recently introduced Philips PU6100 system uses similar methods to those described in Section 6.5.2. A suitable isoeluotropic plane is first established for the four solvents water (or buffer), methanol, THF and acetonitrile, using the methods described earlier. In fact it is not necessary for the plane to be exactly isoeluotropic, and considerable variations in analysis time are acceptable at the three corners of the plane. The system used for retention modelling requires an experimental design of ten chromatographic runs as shown in Fig. 6.5f. One of these ten runs is selected as the reference chromatogram, which provides diode-array spectra of all peaks. These are used in the peak tracking stage, in which the position of each peak is identified in each of the ten chromatograms. Various chemometric methods are used for the deconvolution (unscrambling) of fused or overlapped peaks. This is necessary in order to establish the retention of each individual peak in the fused set (chemometrics means the application of mathematical and statistical methods to chemistry).

The retention of each peak is now modelled across the plane, and from the resulting retention surface the response surface is calculated. The response surface is a three dimensional map showing how chromatographic quality varies with solvent composition at all points on the plane. Various different quality criteria can be used in calculating the response surface; in the example discussed below, involving only two components, the quality function used is simply proportional to the resolution between the peaks.

The results of the optimization are conveniently presented as a triangular contour map, which is a plan view of the response surface showing in the commercial system, regions of different chromatographic quality in different colours. The contour map is provided with a cursor and a simultaneous display of the predicted chromatogram, which changes as the cursor is moved.

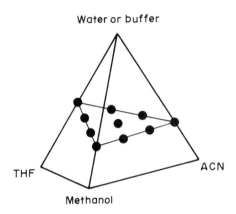

Fig. 6.5f. *Isoeluotropic plane showing position of ten experimental points*

An application of the method has recently been published (R.J. Lynch, S.D. Patterson and R.E.A. Escott, *LC–GC International* 1990, 3, 54). The object was to improve the separation of two positional isomers of an aromatic nitro-compound (the structure of the compound was held to be commercially sensitive). These isomers had previously been separated using a water–acetonitrile gradient as shown in Fig. 6.5g.

Fig. 6.5h shows the chromatograms obtained at the three corners of the isoeluotropic plane that was selected. The variation of retention time obtained is within the acceptable limits of the system. These three chromatograms, together with the seven others obtained at the points shown in Fig. 6.5f, are used to model the retention behaviour and produce the response surface shown in Fig. 6.5i. In this figure the triangular section represents the isoeluotropic plane with the quality function plotted on the vertical axis. The optimum mobile phase composition is the point where the quality function reaches a maximum, which occurs near the THF/water vertex. Fig. 6.5j shows the contour plot obtained, and Fig. 6.5k shows the chromatogram obtained at the predicted optimum mobile phase composition.

∏   What has been gained by the optimization in this example?

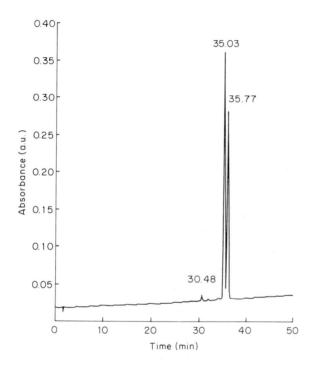

Fig. 6.5g. *Separation of positional isomers using a gradient*

Column: Spherisorb S5 ODS 2, 25 cm × 4.6 mm
Mobile phase: A = 80:20 water/acetonitrile
B = 20:80 water/acetonitrile
gradient: 50–100% B in 50 min
Flow rate: 1 cm³ min⁻¹
Detector: UV absorption, 280 nm

140

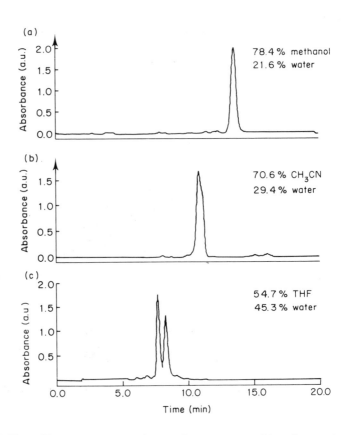

Fig. 6.5h. *Chromatograms obtained at corners of isoeluotropic plane*

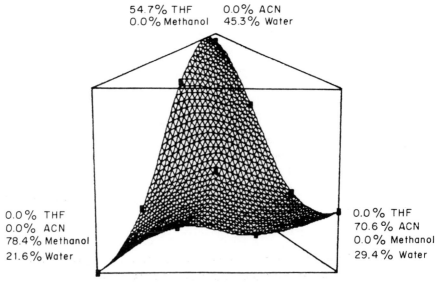

Fig. 6.5i. *Response surface for separation of positional isomers*

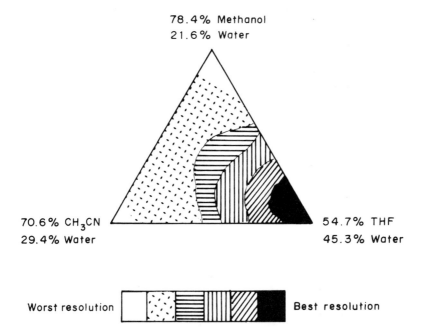

Fig. 6.5j. *Contour plot for separation of positional isomers*

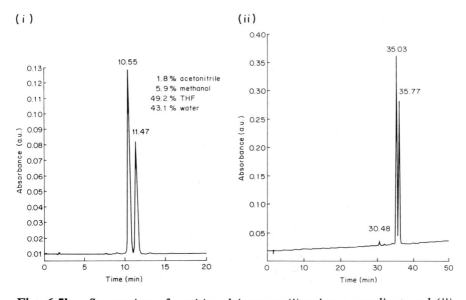

Fig. 6.5k. *Separation of positional isomers (i) using a gradient and (ii) using optimum mobile phase composition*

The resolution of the two isomers is better, and the analysis time has been reduced from about 35 min (+ time for re-equilibration of the column after the gradient) to about 12 min.

The PESOS (Perkin Elmer Solvent Optimization System) again starts with an isoeluotropic plane, which is then searched sequentially; the LC system controller is programmed to perform a set of experiments for the conditions shown by each dot on the triangle in Fig. 6.51, the runs being done in the order shown by the line connecting the dots. Set-up and execution of the experiments is automated, so that the search can take place unattended. The main disadvantage of the method is the time taken, as the system does not stop if an acceptable separation has been found.

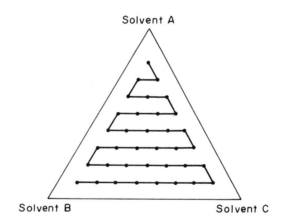

Fig. 6.51. *Experimental design of PESOS*

SAQ 6.5a

In Fig. 6.5m, which is the best chromatogram in each set?

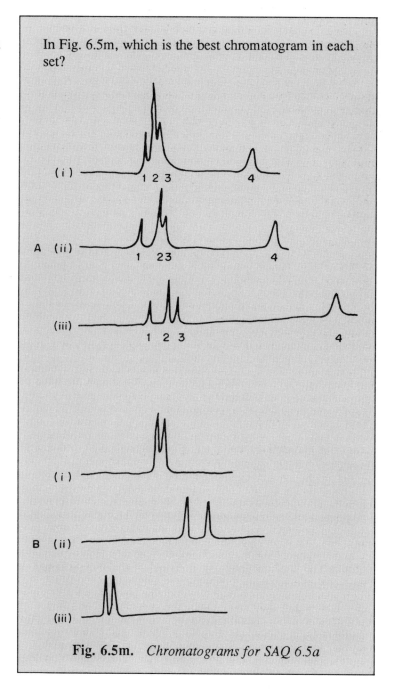

Fig. 6.5m. *Chromatograms for SAQ 6.5a*

SAQ 6.5a

6.6. SUMMARY

Methods are described for the estimation of polarity of solvents and solvent mixtures and for classification of common LC solvents on the basis of selectivity properties. Some of the more important methods of mobile phase optimization are discussed.

Learning Objectives

You should now be able to

- Appreciate the meaning of polarity in HPLC and understand how it is measured;

- Calculate the polarity of mixtures of solvents from the individual solubility parameters;

- Appreciate the classification of solvents on the basis of polarity and selectivity differences;

- Describe in outline some of the methods used for mobile phase optimization in HPLC.

References

POLARITY AND SOLVENT CLASSIFICATION

1. L.R. Synder, *Journal of Chromatographic Science* 1978, **16** 223–234.

2. P.J. Schoenmakers, H.A.H. Billiet and L. deGalan, *Chromatographia* 1982, 15, 205-214.

3. P.A. Sewell and B. Clarke, *ACOL Chromatographic Separations*, John Wiley, 1987, Chapter 3.

MOBILE PHASE OPTIMIZATION

4. L.R. Synder, J.L. Glajch and J.J. Kirkland, *Practical HPLC Method Development*, Wiley–Interscience, 1988.

5. J.C. Berridge, *Techniques for the Automated Optimisation of HPLC Separations*, Wiley–Interscience, 1985.

6. P.J. Schoenmakers, *Optimisation of Chromatographic Selectivity*, Elsevier, 1986, Chapters 2–5.

7. P.J. Schoenmakers, A.C.J.H. Drouen, H.A.H. Billiet and L. deGalan, *Chromatographia*, 1982, **15**, 688–696.

8. J.W. Dolan and L.R. Snyder, *Journal of Chromatographic Science*, 1990. **28**, 379.

9. A. Wright, *Chromatography and Analysis*, April 1990, 5–7.

SAQS AND RESPONSES

SAQ 6.4a — Calculate the composition of binary tetrahydrofuran/water and acetonitrile/water mixtures that are isoeluotropic with methanol/water 50:50.

Response

Using Eq. 6.4a and the δ values from Fig. 6.2a, for methanol/water $50:50$:

$$\delta_m = 0.5 \times 29.4 + 0.5 \times 47.8 = 38.6$$

If Φ_T = required volume fraction of THF:

$$38.6 = \Phi_T \times 18.6 + (1 - \Phi_T) \times 47.8$$

from which

$$\Phi_T = 0.315$$

A neater way to do the calculation is to proceed like this; using Eq. 6.4b, for the isoeluotropic M/W and T/W mixtures:

$$\delta_W - \Phi_M(\delta_M - \delta_W) = \delta_W - \Phi_T(\delta_T - \delta_W)$$

$$\therefore \Phi_T = \Phi_M \left(\frac{\delta_M - \delta_W}{\delta_T - \delta_W} \right) = \left(\frac{29.4 - 47.8}{18.6 - 47.8} \right) \Phi_M$$

$$= 0.63\Phi_M$$

This gives us the T/W composition isoeluotropic with *any* M/W composition (for $\Phi_M = 0.5$, $\Phi_T = 0.315$).

A similar calculation for acetonitrile gives:

$$\Phi_A = \left(\frac{29.4 - 47.8}{23.9 - 47.8} \right) \Phi_M = 0.77\Phi_M$$

so a 38.5% v/v acetonitrile/water mixture has the required polarity.

****

SAQ 6.5a

In Fig. 6.5m, which is the best chromatogram in each set?

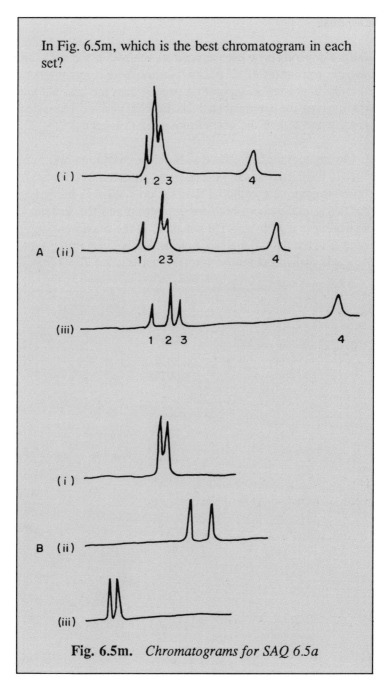

Fig. 6.5m. *Chromatograms for SAQ 6.5a*

Response

A. This depends on what we want to achieve in the separation. If we want baseline resolution of all peaks then the best chromatogram is (*iii*). If we are only interested in separating peak 4 then the best chromatogram is (*i*), which gives the shortest analysis time. If we are interested in peak 1 or peaks 1 and 4 then the best chromatogram is (*ii*).

B. Chromatogram (*iii*) gives us baseline resolution and minimum analysis time. If resolution alone were used as an index of chromatographic quality chromatogram (*ii*) would be the best one. What is needed here is an index that takes into account both the resolution and the analysis time. A simple possibility is $R_S + b(T_a - T)$, where T_a is the maximum acceptable analysis time, T is the actual analysis time and b is a weighting factor. This makes the quality of the chromatogram decrease when $T > T_a$.

****

7. Column Packings and Modes of HPLC

7.1. THE SIZE AND SHAPE OF COLUMN PACKINGS

The materials most commonly used for packings in HPLC columns are microparticulate silicas. These are small porous silica particles with spherical or irregular shape and nominal diameter of 3, 5 or 10 μm. They are manufactured so as to have a narrow particle size and pore size distribution. In bulk, a microparticulate silica resembles fine talcum powder. With microparticulates, dry packing methods result in column beds that are unstable under pressure, so they are packed into columns using a slurry of the material in a suitable solvent and under considerable pressure.

Silica packing material is also available as porous layer beads. These consist of an inert spherical core of glass or plastic, 30–40 μm in diameter, with a thin outer coating of silica or modified silica. These are used nowadays only as material for guard columns (Section 9.3.2) or sometimes for ion exchange separations. Preparative HPLC uses silica particles with 10–20 μm diameter, or larger.

There are limitations to the use of silica, especially at extremes of pH, and to overcome these other packing materials have been developed. These are based on fluorocarbons, carbon, alumina, or polymeric resins. In some areas of HPLC (ion chromatography or exclusion chromatography) the use of polymeric gels or resins is common.

Fig. 7.1a lists some of the properties of two commercially available microparticulate packings. The first is a 3 μm silica bonded with octadecylsilane, the second is a 10 μm unmodified silica.

	Hypersil 3 ODS	Spherisorb S 10W
Supplier	Shandon	Phase Separations
Type	C-18 reverse phase, fully end capped, spherical	unmodified silica, spherical
Nominal diameter, μm	3	10
Range, μm	2.9–3.2	8–12
Pore diameter, μm	1200	800
Range, μm	900–1500	540–1100
Carbon content, %	10	–
Surface area, m^2g^{-1}	170	220
Cost (1990), £ per 10 g	111	43.20

Fig. 7.1a. *Properties of two commercially available microparticulate packings*

You will find as you get further involved with HPLC that there is a quite bewildering variety of microparticulates available in the trade literature (the textbook by Meyer has an extensive list). There is now quite a large number of manufacturers of these materials, and also an increasing tendency to produce 'application-specific' column packings. To keep things in perspective, remember that most of the work in analytical HPLC at present is done with chemically modified silicas, i.e. bonded phases, and of these by far the most important is the nonpolar C-18 type. Even if we restrict ourselves to this one type of packing, there is a large selection available, and although they are all designed to do essentially the same job, there are differences between the C-18 packings from different manufacturers depending on the size, shape and pore size of the silica particle, the carbon content (i.e. the % surface coverage) of the bonded phase and the extent of end-capping. End-capping is a method used to reduce the residual adsorptive properties of the silica, and is discussed in more detail later.

7.2. BONDED PHASES

Silica has silanol (Si–OH) groups which may be chemically modified so as to alter the properties of the silica surface. One way of doing this

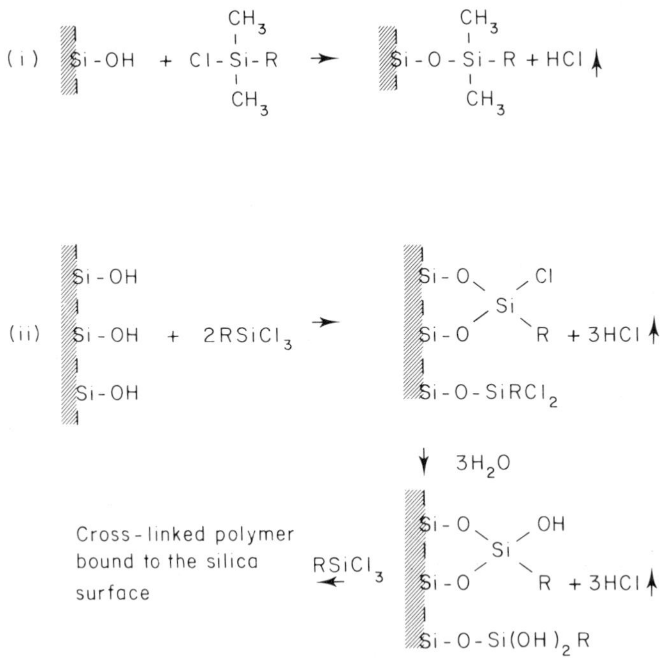

Fig. 7.2a. *Preparation of bonded phases. Reaction of silica with substituted chlorosilanes to form (i) monomeric and (ii) polymeric bonded phases*

is shown in Fig. 7.2a, in which the silica is reacted with a substituted dimethylchlorosilane, with elimination of HCl between a surface silanol group and the silylating agent.

Before reaction, the silica is treated with acid (e.g. refluxed for a few hours with 0.1 mol dm^{-3} HCl). This treatment produces a high concentration of reactive silanol groups at the silica surface, and also removes metal contamination and fines from the pores of the material. After drying, the silica is then refluxed with the substituted dimethylchlorosilane in a suitable solvent, washed free of unreacted silane and dried. This reaction produces what is called a 'monomeric' bonded phase, as each molecule of the silylating agent can react with only one silanol group.

More complicated surface structures can be produced by changing the functionality of the silylating agent and the conditions under which the reaction is carried out. The use of di- or tri-chlorosilanes in the presence of moisture can produce a cross-linked polymeric layer at the silica surface, as shown in Fig. 7.2a(ii). Monomeric bonded phases are preferred, as their structure is better defined and they are easier to manufacture reproducibly than the polymeric materials.

Different kinds of bonded phases can be made by varying the nature of the functional group R in the silylating agent. For example, a bonded phase cation exchanger can be made by using a phenyl or phenyl-substituted alkyl group. After the bonding reaction, the phenyl is sulphonated using chlorosulphonic acid. An anion exchanger can be made by using a chlorinated alkyl group, which then forms a quaternary salt by reaction with a tertiary amine.

Many other methods have been used to prepare bonded phases; these include esterification of the surface silanol groups with alcohols, or conversion of the silanol –OH to –Cl using thionyl chloride, followed by reaction with an amine or an organometallic compound. If you are interested, there are details in the textbooks by Knox and by Hamilton and Sewell. Since it is not possible to bond all of the surface silanol groups, unreacted silanols capable of adsorbing polar molecules will affect the chromatographic properties of the bonded phase. Usually, in reverse phase separations, interaction of the silanols with polar groups causes undesirable effects such as tailing, poor resolution and excessive retention. The concentration of unreacted silanols in non-polar bonded phases is usually reduced by the process of end-capping, in which most of the remaining silanols are reacted with a small silylating agent, such as trimethylchlorosilane. Fig. 7.2b shows the

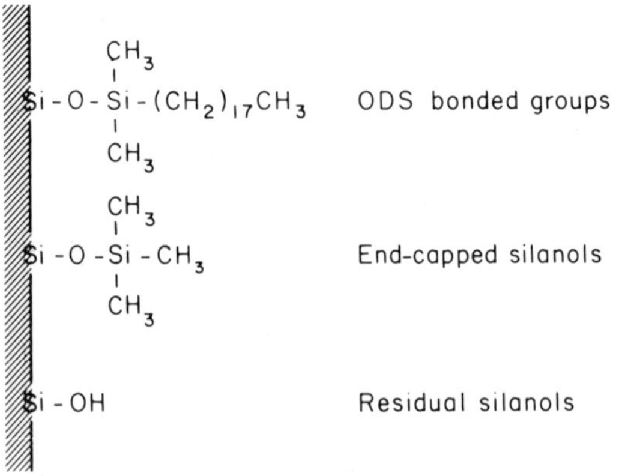

Fig. 7.2b.   *Structure of an ODS (C-18) surface*

surface structure of an octadecylsilane bonded phase, containing bonded C-18 alkyl groups, end-capped silanol groups and a small number of free silanols.

Fig. 7.2c shows a small selection of commercially available bonded phases. Most chromatography suppliers give comprehensive lists of these.

Name	Attached functional group	Type	Particle shape and size, μm	Price, £ per 10g (1990)	Manufacturer (supplier)
Hypersil SAS	C-1	NP	Spherical, 5	99.84	Shandon Southern
Spherisorb S5P	Phenyl	NP	Spherical, 5	89	Phase separations
Lichrosorb RP-8	C-8	NP	Irregular, 10	119.50	E. Merck (Hichrom)
Nova-Pak C-18	C-18	NP	Spherical, 4	(i)	Waters
Nucleosil 3C-18	C-18	NP	Spherical, 3	119.60	Macherey Nagel (FSA)
Partisil 10 SAX	quaternary ammonium	SAX	Irregular, 10	114.87	Whatman
Partisil 10 SCX	sulphonic acid	SCX	Irregular, 10	144.87	Whatman
Lichrosorb CN	$-(CH_2)_3CN$	WP	Irregular, 10	119.50	E. Merck
Spherisorb S3NH$_2$	$-(CH_2)_3NH_2$	P or WAX	Spherical, 3	97.60	Phase separations

(i) Available only as packed columns.
NP, WP, P = non-polar, weakly polar, polar.
SCX, SAX, WAX = strong cation, strong anion, weak anion exchanger.

Fig. 7.2c

Most sorts of packing material can be bought loose, either for repairing columns or for making them. Although it is not difficult to make columns, most HPLC users buy columns from manufacturers. Fig. 7.2d gives a brief description and current prices of a few manufactured columns. These are two general purpose C-18 modified silicas, a wide pore specialty protein column, an chiral column, a resin-based ion exchanger and an exclusion column.

Name	Description	Length (cm) × id (mm)	Price, £ (1990)	Supplier
Nova-pak C-18	General purpose reverse phase. 4 μm spherical C-18 modified silica	15 × 3.9	152.25	Waters
Spherisorb S5 ODS-1	As above, but 5 μm	15 × 4.6	96	Phase Separations
Vydac C8	Speciality protein. 5 μm C-8 modified wide pore silica	15 × 4.6	260	Technicol
Resolvosil BSA 7	Chiral, based on bovine serum albumin bonded to wide pore Nucleosil	15 × 4.0	345	Thames Chroma- tography
IC-Pak A	Polymethacrylate anion exchanger	5 × 4.6	636.30	Waters
PL-gel 5 50 Å	Polystyrene–DVB GPC column	30 × 7.5	619.61	FSA

Fig. 7.2d

7.3.   OTHER PACKING MATERIALS

Styrene-divinylbenzene resins can be used for reverse phase chromatography as well as for ion exchange or exclusion applications. They are produced by copolymerizing styrene and divinylbenzene as shown in Fig. 7.3a. The amount of divinylbenzene used determines the degree of cross-linking and thus the pore structure obtained. A high degree of cross-linking produces a rigid gel which is stable at high pressures and does not swell in contact with solvents. The resin can be modified by the introduction of suitable functional groups into the structure, e.g. C-18 for reverse phase applications or sulphonic acid or quaternary ammonium for ion exchange. Styrene–divinylbenzene resins have much better pH stability than silica in aqueous systems.

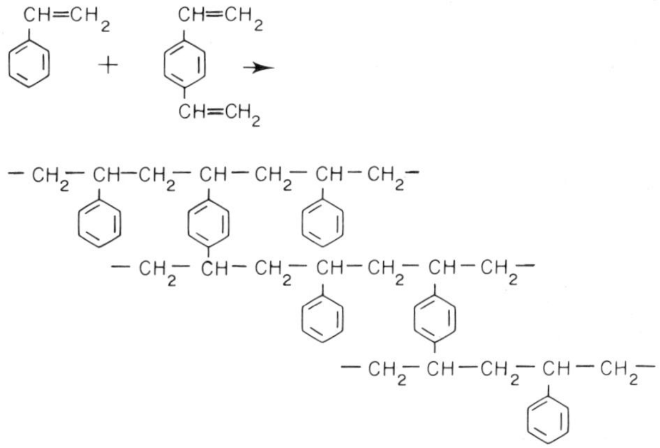

Fig. 7.3a. *Styrene–divinylbenzene resin*

The limitations of silica at high pH have prompted much work on the development of alternative materials for use under such conditions. Bonded silica packings that are stable from pH 2–13 have recently become available, with C-1 and C-18 functionalities. Hypercarb (a porous graphitic carbon, introduced by Shandon in 1988) is a reverse phase material that can be used throughout the whole pH range. It is useful for the separation of basic solutes and also shows promise for the separation of chiral compounds.

Unisphere PBD (Biotage Inc.) is a reverse phase packing consisting of spherical polybutadiene-coated alumina particles. The pH range is 2–13 and the coating process provides a highly deactivated surface. This allows many separations to be done without the use of mobile phase additives, which are often needed to reduce adsorptive effects with silica packings. In addition, the material has significantly better flow properties and higher sample capacity than microparticulate silica. Column pressure drops are typically 25–50% of those obtained with spherical silicas, thus flow rates can be increased, resulting in shorter analysis times. Scaling up to preparative separations is also simplified. The packing is currently available as 3 or 8 μm particles.

7.4. MODES OF HPLC

Microparticulate silicas have been used in a number of different ways in HPLC (these were mentioned in the introduction):

(*a*) As adsorbents;

(*b*) As supports for stationary liquids in partition chromatography;

(*c*) As bonded phases;

(*d*) As materials for exclusion chromatography.

Methods (*a*) and (*b*) were the original modes of HPLC but have been replaced to a large extent by the use of bonded phases. Liquid–liquid separations, in particular, are now very seldom performed. Chromatography using bonded phases is easier, more versatile and quicker than the older modes, and has much better reproducibility. In a bonded phase, the highly polar surface of the silica is altered by the attached functional groups, which can be non-polar (e.g. C-18, phenyl, C-8) polar ($-NH_2$, $-CN$) or ionizable (sulphonic acid, quaternary ammonium). The introduction of ionizable groups produces bonded phases with ion exchange properties.

Fig. 7.4a is an indication of how an HPLC method might be selected based on the molecular mass and the solubility of the sample. You can see that for many samples there is a choice of method, but that in many cases separation can be achieved by reverse phase chromatography using a bonded silica stationary phase. This is the mode that we would tend to look at first; it is often faster, cheaper and experimentally easier than the alternatives.

7.5. NORMAL AND REVERSE PHASE CHROMATOGRAPHY

The terms normal phase and reverse phase are used to describe adsorption and many bonded phase separations (but not in connection with ion exchange or exclusion). Normal phase means that the polarity of the stationary phase is higher than that of the mobile phase, which is what happens, for example, when silica is used in adsorption chromatography. Reverse phase means that the polarity of the stationary phase is less than that of the mobile phase, which is the case with hydrocarbon-type bonded phases and polar mobile phases. Polar bonded phases can be used in either normal or reverse phase modes. With both techniques solutes are eluted in order of polarity, with reverse phase most polar first and with normal phase least polar first, and we can change the retention times of solutes by changing the polarity of the stationary phase or (more easily) of the mobile phase. These facts are summarized in Fig. 7.5a. For ionizable solutes, the pH of the mobile phase is an important factor in the control of retention and selectivity.

Reverse phase operation with bonded phases has achieved very wide popularity because it has the following advantages:

(a) The method has a very broad scope that allows samples with wide ranges of polarity to be separated. There is the possibility of using many different bonded phases, producing a very flexible separating system;

(b) The method uses relatively inexpensive mobile phases, and equilibration of the mobile phase with the column is rapid;

(c) It can be applied to the separation of ionic or ionizable compounds by the use of ion pairing or ion suppression techniques (see Section 7.6.7);

(d) The mode is generally experimentally easier, faster and more reproducible than other HPLC modes.

Reverse phase operation is, however, not without limitations. The following are among the more important of these.

(a) For many silica bonded phases, stable columns can only be maintained over a pH range between about 3 and 8. Below pH 3 the bonded group may be removed, and above pH 8 the silica is appreciably soluble in the mobile phase. For separations outside these pH limits the materials mentioned in Section 7.3 can be used.

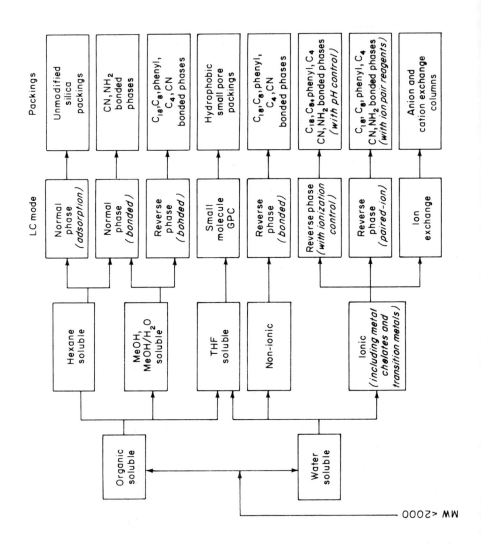

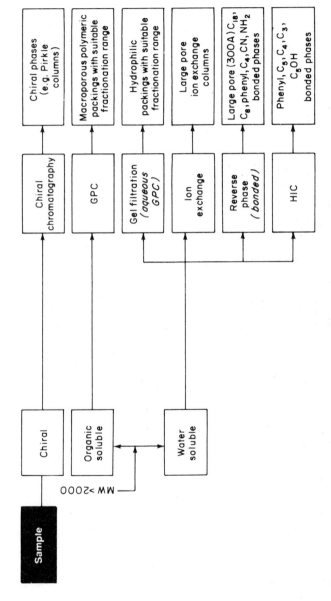

Fig. 7.4a. *Choice of HPLC method*

	Normal phase	Reverse phase
Stationary phase polarity	high	low
Mobile phase polarity	low–medium	medium–high
Typical mobile phase	heptane/$CHCl_3$	CH_3OH/H_2O
Order of elution	least polar first	most polar first
To increase retention of solutes	decrease mobile phase polarity	increase mobile phase polarity

' **Fig. 7.5a.** *Characteristics of normal and reverse phase chromatography*

(*b*) The presence of unreacted silanol groups on the silica surface can often cause tailing, excessive retention time and non-reproducible behaviour between columns due to solute adsorption. These effects can in many cases be overcome, as will be seen later.

(*c*) The reverse phase retention mechanism is still not properly understood. One possible explanation is that the hydrophobic surface of the bonded phase extracts the less polar constituent of the mobile phase to form a layer at the silica surface and that partitioning of solutes then occurs between this layer and the mobile phase. However, in many cases in reverse phase chromatography there may be several mechanisms operating at the same time (adsorption on unreacted silanol groups, for instance). On the basis that a better understanding of separation mechanisms with bonded phases will expand their uses, there is much study currently directed towards this problem.

Solute retention in reverse phase separations is governed by a number of factors, the more important of which were listed in Fig. 7.5a. Solutes are generally eluted in order of polarity, the most polar being eluted first. If we think of a reverse phase separation as being a partitioning process in which solutes are distributed between a non-polar stationary phase and a polar mobile phase, then the non-polar solutes will be soluble in the stationary phase and will travel through the system more slowly than the polar solutes, which favour the mobile phase. We can change the distribution by changing the polarity of the mobile phase. For instance, if we make the mobile phase less polar (by increasing the ratio of organic solvent to water) then we will shift the distribution of solutes towards the mobile phase, and their retention will decrease. we can also alter retention by changing the polarity of the stationary phase, which depends on the type of non-polar group used (for instance, phenyl > C-8 > C-18) and on the percentage of carbon present.

∏   What would be the effect on the retention of a given solute of:

(*a*) Changing from a C-18 to a phenyl bonded phase?

(*b*) Changing from a C-18 bonded phase to another C-18 that contained a smaller percentage coverage of the hydrocarbon?

(*a*) As the phenyl bonded phase is more polar than the C-18 we would expect retention to decrease.

(*b*) Similarly, decreasing the percentage of carbon in a given type of bonded phase will increase the polarity, so that, other things being equal, we would expect retention to decrease here as well. In commercial C-18 packings the carbon content of the bonded phase can be between 5 and 30%. Fig. 7.5b

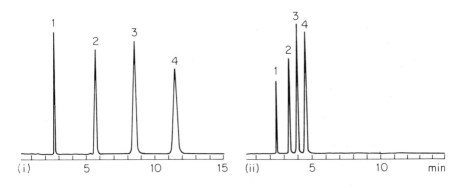

Fig. 7.5b. *Reverse phase chromatogram of test mixture*

Column:	(*i*) 4 μm C-18
	(*ii*) 4 μm –CN, both 15 cm × 3.9 mm
Mobile phase:	CH_3CN/H_2O 40:60
Flow rate:	2 cm^3 min^{-1}
Sample:	1 Benzyl alcohol
	2 2-Phenoxyethanol
	3 4-Methoxybenzaldehyde
	4 Methyl phenyl ether
Detector:	UV absorption, 254 nm

is a chromatogram of a test mixture on (*i*) a C-18 and (*ii*) a –CN bonded phase that can be used either in normal or reverse phase mode. In each case the solute with the highest polarity (the alcohol) is eluted first and the solute with the lowest polarity (the ether) is eluted last. Increasing the polarity of the bonded phase on going from C-18 to –CN causes the retention of all solutes to decrease.

Fig. 7.5c shows a reverse phase separation of some tricyclic antidepressant drugs. These compounds are weak bases, and at the pH used will be completely protonated (see Section 7.6.3.). Because the protonated bases are very polar compounds, they are adsorbed strongly by unreacted silanol groups, causing excessive retention and severe peak tailing. Effects like these are possible even with end-capped bonded phases. There are several ways to approach the problem. One is to add to the mobile phase a large concentration (relative to the concentration of the solutes) of a competing base. Because of its relatively high concentration it is preferentially adsorbed by the silanol groups, thus minimizing the adsorption of the other bases. The figure shows the dramatic change that results when nonylamine is added as a competing base. Other possibilities for this separation would be to use ion suppression or ion pairing techniques (Section 7.6.7).

Operation of bonded phases in the normal phase mode is now replacing many separations that were previously done by adsorption chromatography with unmodified silica. Compared to silica, the bonded phases show less tailing and respond more rapidly to changes in mobile phase composition. The chromatography is generally more reproducible. They can also show different selectivities, depending on the nature of the polar bonded group. Weakly polar bonded phases include diol, cyano or nitro groups; more polar types contain amino groups. In fact, polar bonded phases are most often used in the reverse phase mode, with polar solvents. Amine bonded phases are especially useful for the reverse phase separation of carbohydrates (Fig. 7.5d shows a typical example) and can also be used as weak anion exchangers, for instance in the separation of organic acids.

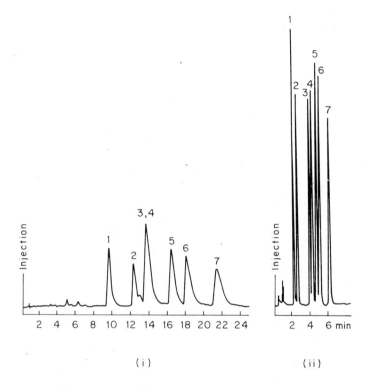

Fig. 7.5c. *Separation of antidepressant drugs*

Column: C-8 bonded phase 15 cm × 4.6 mm
Mobile phase: (*i*) $CH_3CN/0.01$ mol dm^{-3} H_3PO_4 adjusted to pH 2.5 with KOH
               (*ii*) $CH_3CN/0.01$ mol dm^{-3} H_3PO_4 + 0.005 mol dm^{-3} nonylamine, pH 2.5
Flow rate: 2 cm^3 min^{-1}
Detector: UV absorption, 254 nm
Sample: 1 Nordoxepin   2 Doxepin   3 Desipramine
               4 Protriptyline   5 Imipramine   6 Nortriptyline
               7 Amitriptyline

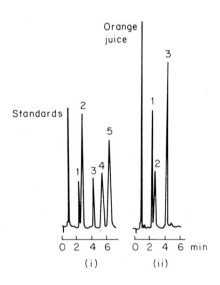

Fig. 7.5d. *Separation of carbohydrates*

Column: 5 μm $-NH_2$ bonded phase, 25 cm $\times$ 4.6 mm
Mobile phase: CH_3CN/H_2O 78 : 22
Flow rate: 2 cm^3 min^{-1}
Detector: Refractive index
Chromatograms: (*i*) Standards
                   (*ii*) Orange Juice diluted in two parts of CH_3CN
Peaks: 1 Fructose   2 Glucose   3 Sucrose
                   4 Maltose   5 Lactose

SAQ 7.5a

A test mixture consisting of phenyl methyl ketone, nitrobenzene, benzene and methylbenzene is to be separated on a C-18 column with a mobile phase of CH_3OH/H_2O 60 : 40. With these conditions, the ketone is eluted first.

(*i*) In what order are the other solutes eluted?

(*ii*) How would you change the composition of the mobile phase so as to increase the retention of the solutes?

(*iii*) How would the retention of the solutes be affected by using a phenyl bonded phase instead of the C-18?

(*iv*) If the C-18 bonded phase contained unreacted silanol groups, how would the retention of the solutes be affected by end-capping the stationary phase?

7.5.1. Hydrophobic Interaction Chromatography (HIC)

In the reverse phase chromatography of proteins and peptides it is sometimes necessary to avoid the use of organic solvents in the mobile phase,
as these cause denaturing of the sample. With conventional reverse phase
columns, the use of organic solvents is necessary to prevent excessively long
retention times. HIC uses a stationary phase that is only weakly hydrophobic
(i.e. is slightly more polar than a conventional reverse phase packing). This
is achieved by the use of a bonded phase with a short carbon chain, such
as methyl, propyl or phenyl. Proteins are separated by using a descending
salt gradient, running from high to low salt concentrations. Fig. 7.5e shows
an example of such a separation.

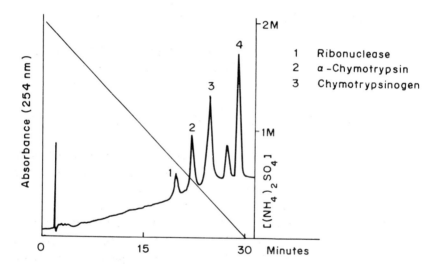

Fig. 7.5e. *Separation of protein mixture by HIC*

Column: Synchropak Propyl 7.8 cm × 7.5mm
Mobile phase: A: 2 mol dm^{-3} $(NH_4)_2SO_4$
                        0.1 mol dm^{-3} KH_2PO_4 pH 6.8
                      B: 0.1 mol dm^{-3} KH_2PO_4 pH 6.8
Flow rate: 1 cm^3 min^{-1}
Detector: UV absorption, 254 nm

7.5.2. Summary

The surface of microparticulate silica can be modified by the attachment of different groups to produce bonded phases. Reverse phase chromatography using bonded phases is generally faster and easier than other modes, and consequently has achieved very wide popularity.

Learning Objectives

You should now be able to:

• Specify types of bonded phases that are used for various applications;

• Describe some methods used for the preparation of bonded phases;

• Predict how solute retention times on bonded phases will be affected by a change in the polarity of the stationary or mobile phase.

7.6. THE CHROMATOGRAPHY OF IONIC SOLUTES

Separation of ionic solutes is now an important and rapidly growing field. Ion chromatography or IC is the term that is currently used to describe separations of this type. There has been a tendency in the past, actively encouraged by some manufacturers, to regard IC as a technique separate to and distinct from HPLC, which it is not. The purpose of this deception is to persuade us that if we want to separate ions we have to buy an expensive ion chromatograph, whereas such separations can frequently be done on existing instruments.

Originally, IC was done using as stationary phases materials very similar to classical ion exchange resins, with detection by conductance measurement. Such systems are still widely used; alternatively we can use ion exchangers based on modified silica, or often ions can be separated using reverse phase techniques. Conductance measurement is not the best detection method for some ions; most manufacturers now suggest that a range of detection systems should be used.

7.6.1. Ion Exchange Chromatography

An ion exchange resin consists of an insoluble, rigid three-dimensional matrix, for example polystyrene cross-linked with a small amount of divinylbenzene to produce mechanical stability. The structure of this was shown in Fig. 7.3a. The surface of this matrix contains ionizable sites (e.g. sulphonic acid or quaternary ammonium) that can carry a positive or a negative charge. Each of these sites also requires an oppositely charged ion (the counter ion) for overall neutrality. If the ionizable sites are positively charged, the counter ion is an anion and the resin will exchange anions from solution:

$$R^+Y^- \ + \ X^- \ \rightleftharpoons \ R^+X^- \ + \ Y^-$$
$$\text{(resin)} \quad \text{(solution)} \qquad \text{(resin)} \quad \text{(solution)}$$

In this scheme, a counter ion Y^- attached to the R^+ site on the resin is exchanged for another anion X^- from solution. To exchange cations we need a structure with exchange sites that are negatively charged and associated positive counter ions:

$$R^-Y^+ \ + \ X^+ \ \rightleftharpoons \ R^-X^+ \ + \ Y^+$$
$$\text{(resin)} \quad \text{(solution)} \qquad \text{(resin)} \quad \text{(solution)}$$

A mixture of ions can be separated if we have different strengths of interaction between the solute ions in the mobile phase and the fixed exchanging groups on the stationary phase.

The terms strong or weak as applied to ion exchange resins indicate how the exchanging properties of the structure vary with pH. A strong cation exchanger (SCX) contains sulphonic acid groups that are fully ionized above about pH 2. A strong anion exchange resin (SAX) contains quaternary ammonium exchange sites that are fully ionized up to about pH 10. Weak cation and anion exchangers (WCX or WAX) contain, respectively, carboxylic acid and amino groups, which are ionized only over a restricted range of pH. The capacity of a resin measures the amount of material that can be exchanged by a given amount of the resin. Variation of capacity with pH for the various types is shown in Fig. 7.6a; note that WCX and WAX resins have a higher capacity than SCX or SAX types.

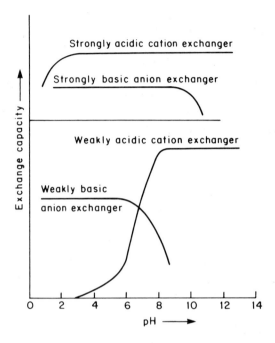

Fig. 7.6a. *Variation of ion exchange capacity with pH*

In addition to the resins described above, other ion exchange materials available for HPLC are:

(*a*) Porous layer beads, consisting of a solid core of glass or polymer with a thin surface layer of ion exchange material, or a silica bead with ion exchange groups bonded to the surface;

(*b*) Bonded phases based on microparticulate silica.

Fig. 7.6b summarizes the main differences between these materials. Porous layer beads are nowadays very rarely used in analytical columns, although they are still used in guard columns. The textbook by Meyer has an extensive list of commercially available materials.

	Styrene–DVB	Porous layer beads	Bonded silica
Particle size, μm	5–20	30–50	5–10
Capacity	high	low	high
Sample loading	large	small	moderate
pH range	2–14	2–9	2–8
Packing method	slurry	dry	slurry
Efficiency	low ———————————————————→ high		

Fig. 7.6b. *Ion exchange materials used for HPLC*

7.6.2. Separation Modes in IC

Inorganic or organic ionic or ionizable solutes can be separated using a variety of strategies, which are summarized in Fig. 7.6c. We can separate the solutes as ions using an ion exchange resin or bonded silica, or it may be possible to convert the ions to polar molecules, after which they can be separated on a reverse phase system. Ion exclusion is a recent development which is useful in one or two specialized separations, e.g. mixtures of weak acids. Separation here is by the size exclusion mechanism (Section 7.7.2). Ions of the same sign as the ionic functional groups of the ion exchanger are repelled and prevented from penetrating the stationary phase pores.

7.6.3. Factors Affecting Retention

Retention of solutes in ion exchange chromatography is determined by the nature of the sample, the type and concentration of other ions present in the mobile phase, pH, temperature and the presence of solvents. Because there are so many variables, it is often not easy to predict what will happen in an ion exchange separation if we change the experimental conditions. There are some useful guidelines, and to see how they work we will consider the chromatography of a weak acid, e.g. benzoic acid.

	Ion exchange	Ion pair	Ion suppression	Ion exclusion
Column type	anion or cation exchange resin, or modified silica 5 cm–25 cm	silica C-18 12.5 cm–25 cm	silica C-18 12.5–25 cm	sulphonated resin 30 cm × 5 cm
Mobile phase	buffer or buffer + organic modifier	buffer + ion pair reagent or buffer + organic modifier + ion pair reagent	buffer + organic modifier	dilute acid
pH range	resins: 2–12 silica: 2–8	2–8	2–8	about 3
To change retention	pH or buffer type or buffer concentration or nature or concentration of organic modifier	pH or nature/concentration of pair reagent or nature/concentration of modifier or buffer type/concentration	pH or nature/concentration of buffer or organic modifier	little change possible

Fig. 7.6c. *Separation modes in IC*

$$C_6H_5COOH \rightleftharpoons C_6H_5COO^- + H^+$$

$$K_a = \frac{[H^+][C_6H_5COO^-]}{[C_6H_5COOH]} = 6.3 \times 10^{-5}$$

$$pK_a = -\log_{10} K_a = 4.2$$

Fig. 7.6d. *Ionization of benzoic acid*

In solution, the acid is partly dissociated, i.e. it is present both as benzoate anions and benzoic acid molecules. The pK_a value measures the strength of the weak acid and hence its degree of dissociation; the higher the pK_a value the weaker the acid.

∏   What would be the effect on the equilibrium above if the benzoic acid was made up in (*a*) an acid buffer (*b*) an alkaline buffer?

Adding a common ion (e.g. H^+) shifts the equilibrium to the left, so if the buffer pH is low enough the acid would be present only as C_6H_5COOH molecules. Similarly, we can force the acid to ionize completely if the pH is high enough. As a rough guide, ionization is suppressed at $pK_a - 1$ and is complete at $pK_a + 1$.

In an ethanoate buffer at around pH 5.5 we can assume that the benzoic acid is present as $C_6H_5COO^-$. The acid is retained on an ion exchange column because of competition for the stationary phase exchange sites between the ethanoate anions in the mobile phase and the benzoate anions in the sample.

$$R^{+-}OOCCH_3 + {}^-OOCC_6H_5 \rightleftharpoons R^{+-}OOCC_6H_5 + {}^-OOCCH_3$$
         (resin)   (solution)   (resin)   (solution)

∏   If the concentration of ethanoate in the buffer was increased, keeping the pH the same, how would the retention of the benzoate change?

Increasing the ethanoate concentration in the solution forces the equilibrium above to the left. The concentration of benzoate in the mobile phase will increase, so its retention will decrease. Similarly, we could increase retention by decreasing the ethanoate concentration.

Change in the pH of the solution will alter the proportions of C_6H_5COOH and $C_6H_5COO^-$ in the sample. Decreasing the pH will reduce the concentration of benzoate ion and so reduce the retention. The presence of appreciable amounts of both forms is likely to cause tailing. A mixture of weak acids will elute in order of decreasing pK_a; the stronger acids (smaller pK_a) will be more ionized and so will be retained longer by the stationary phase. In general, polyvalent ions are retained longer than divalent, divalent longer than monovalent. Of the same charge type, smaller ions are generally retained longer than larger ions. These general rules apply to simple inorganic ions as well.

∏ If the ethanoate in the buffer was replaced by citrate and there were no other competing equilibria what would be the effect on retention?

The citrate would be strongly retained by the stationary phase, so that the retention of the acid would decrease. However, if mobile phase buffers are formed using polyvalent salts there is a strong possibility of complex formation, which will alter the predicted behaviour of the system.

Ion exchange equilibria are usually established faster at higher temperatures. Increasing the temperature will improve efficiency, decrease retention and may alter the selectivity of the separation. The use of organic solvents as mobile phase modifiers generally causes retention to decrease, but because the use of organic solvents will change many of the variables in an ion exchange separation their effects are not easy to predict.

7.6.4. Precipitation and Complexation

You will have seen from the preceding paragraphs that ionizable solutes can exist in solution in various forms, not always as ions. Sulphate, for example, at most pH values is present in solution as SO_4^{2-} but low pH will exist as HSO_4^- and at extremely low pH as H_2SO_4. This is a consequence of the equilibrium:

$$H_2SO_4 \rightleftharpoons HSO_4^- + H^+ \rightleftharpoons SO_4^{2-} + H^+$$
$$pK_{a1} < 0 \qquad\qquad pK_{a2} = 1.99$$

Thus above pH 3 sulphate is present predominantly as SO_4^{2-}, below pH 1 as HSO_4^-. The pK represents the pH at which equal amounts of both forms are present. Thus $[SO_4^{2-}] = [HSO_4^-]$ when pH = $pK_{a2} = 1.99$.

Cations in solution may precipitate out in the presence of various anions. For a sparingly soluble solid M_aL_b in equilibrium with its ions in aqueous solution:

$$M_aL_b \rightleftharpoons aM^{b+} + bL^{a-}$$

the solubility product can be expressed in terms of the molar concentrations of the ions:

$$K_{sp} = [M^{b+}]^a \cdot [L^{a-}]^b$$

For example, given that $K_{sp}(AgCl) = 2 \times 10^{-10}$ mol^2dm^{-6}, if we mix equal volumes of 2×10^{-3} mol dm^{-3} $AgNO_3$ and HCl then in the resulting solution $[Cl^-] = 1 \times 10^{-3}$ and $[Ag^+] = 2 \times 10^{-7}$ mol dm^{-3}—almost all of the silver is now present as silver chloride. For chromatography, it is important to know the form of our ionizable solutes in solution, and to this end we can make useful predictions by using pK_a values (for anions) or K_{sp} values (for cations).

SAQ 7.6a

For these problems you need to know the pK_a values given below:

	pK_{a1}	pK_{a2}	pK_{a3}
H_3PO_4	2.13	7.20	12.36
H_2CO_3	6.35	10.33	
HF	3.17		
HCl	<1		

(*i*) In what form or forms is phosphate present at pH 1, at pH 2 and at pH 9? Other things being equal, how would the retention of phosphate change with pH on an anion exchange column?

$\longrightarrow$

SAQ 7.6a
(cont.)

(*ii*) You are separating carbonate and chloride using an anion exchange column and a pH 8.5 buffer as the mobile phase. With this system, carbonate elutes just before chloride. How would you expect the retention of these two to alter if you change the mobile phase to 5×10^{-3} mol dm^{-3} KOH?

(*iii*) Why would it be unwise to run fluoride on an anion exchange column at pH 2.5?

7.6.5. Detection Modes in IC

The modes commonly used are conductivity, UV absorption, refractive index or electrochemical detection. Conductivity detection is a good general purpose method; the others usually exhibit high sensitivity for a reduced number of ions and so tend to be used as selective detection systems.

Both conductivity and UV absorption may be used in direct or indirect mode, corresponding to an increase or decrease, respectively, of the measured property as the solute band is eluted. For conductivity detection the baseline conductance reading is due to the various ions in the mobile phase buffer. When a band of solute ions elutes (increasing the conductance) the solute ions displace the mobile phase ions in this region (decreasing the conductance). The net result will depend on the relative magnitude of these two effects, as shown in Fig. 7.6e.

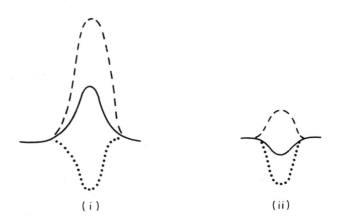

(i)   (ii)

Fig. 7.6e. *Direct (i) and indirect (ii) conductivity detection*

- - - -, conductivity change due to solute ions
......, conductivity change due to removal of mobile phase ions
_____, net conductivity change

∏   Many ion chromatograms show a 'water dip' (a negative peak at the position on the chromatogram corresponding to an unretained solute). Can you explain the reason for this effect, using a similar argument to the one above?

If we inject a solution of ions in water, the water will be unretained on the column and a band of relatively highly conducting mobile phase will be replaced by a band of very weakly conducting water, causing a large negative peak.

In direct UV detection we have a UV transparent mobile phase and the solute ions absorb UV. Indirect UV detection uses a UV absorbing mobile phase with solute ions that do not absorb, thus producing a decrease in absorbance as the solute band passes through the detector.

A number of ions, e.g. Br^-, NO_3^-, NO_2^-, SCN^- and I^-, absorb UV radiation between 200 and 220 nm and so can be detected directly. The main advantage of this method is that it does not detect a number of common ions, Cl^- for example. This allows us to detect ions like NO_3^- or NO_2^- in the presence of a large Cl^- excess. In some cases, UV detection can be used after derivatization. The most common example of this is 4-(2-pyridylazo) resorcinol, or PAR, derivatization. The PAR complexes with transition metals absorb strongly at around 520 nm.

Electrochemical detection is used for low level or selective analysis of the three important anions I^-, SO_3^{2-} and CN^-. Retractive index detection is used mainly for borate and polyphosphonates.

7.6.6. Techniques Used with Conductivity Detectors

These detectors measure the conductance (1/resistance) rather than the conductivity (1/resistivity) of the eluent, although they are commonly called conductivity detectors. One of the main problems of conductivity detection with IC is that the mobile phase used must contain ions and may have a conductance that is high compared with the conductance due to the solute ions, so with conductivity detection we would be trying to measure a small change in a large quantity, which at first sight might seem to be unpromising.

∏   Why must the mobile phase contain ions?

The solute ions and the mobile phase ions compete for sites on the ion exchange resin surface. If there were no ions in the mobile phase, the solute ions would not be removed from the surface of the resin.

The solution is to compensate for the mobile phase conductance. Two approaches to this have been used, known as suppressed IC (used by Dionex) and unsuppressed IC (used by Waters and others). The principle of suppressed IC is that, after flowing through the ion exchange column, the eluent passes through another ion exchange system that converts mobile phase ions into non-ionized or very weakly ionized species, whilst the solute ions are unaffected.

As an example, consider the separation of anions using a sodium carbonate/sodium bicarbonate buffer as the mobile phase. After passing through the anion exchange column the eluent flows through a cation exchange column in the hydrogen form (the suppressor column). The sodium ions in the buffer are exchanged for H^+, converting the buffer to weakly ionized H_2CO_3.

$$R–SO_3^-H^+ + Na^+ \rightleftharpoons R–SO_3^-Na^+ + H^+$$
$$\longrightarrow$$
$$H^+ + HCO_3^- \rightleftharpoons H_2CO_3$$

∏   Can you think of a similar arrangement that could be used for cations?

The cations could be separated in dilute HCl and subsequently pass through a strong anion exchanger in the OH^- form; this would convert the HCl mobile phase to H_2O.

The use of ion exchange columns in this way for suppression increases the dispersion of the system, and also the suppressor columns have to be regenerated from time to time. In practice, rather than using conventional ion exchange columns, suppression is done using devices called membrane suppressors, which have a small internal volume and do not require any regeneration.

Unsuppressed IC uses relatively low conductivity mobile phases and a detector in which the suppression of the mobile phase conductivity is done electronically. Suppressed IC is the older of the two methods and has the advantage that a great deal of work has been published using it. It is, however, relatively expensive and complex. Unsuppressed IC is in practice a simpler method that is easier to use, and a wider variety of eluents is available.

Fig. 7.6f shows an anion separation using unsuppressed conductivity detection. Fig. 7.6g is a cation separation using UV detection after derivatization with PAR. After separation, the column eluent is combined with PAR in a mixing tee followed by a 2m × 0.5 mm i.d. reaction coil before the detector.

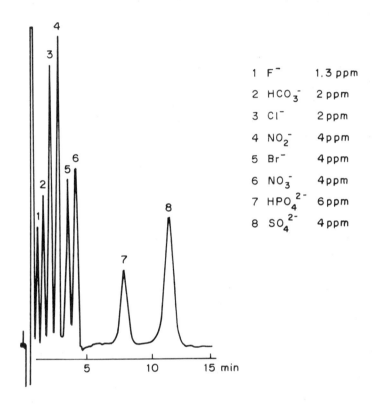

1	F^-	1.3 ppm
2	HCO_3^-	2 ppm
3	Cl^-	2 ppm
4	NO_2^-	4 ppm
5	Br^-	4 ppm
6	NO_3^-	4 ppm
7	HPO_4^{2-}	6 ppm
8	SO_4^{2-}	4 ppm

Fig. 7.6f. *Anion separation by ion exchange*

Column: Waters IC-PAK A, 5 cm × 4.6 mm
Mobile phase: 1.3 mmol dm^{-3} potassium gluconate/1.3 mmol dm^{-3} borax pH 8.5
Detector: Conductivity

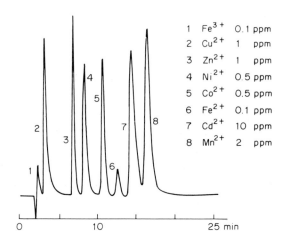

1	Fe^{3+}	0.1 ppm
2	Cu^{2+}	1 ppm
3	Zn^{2+}	1 ppm
4	Ni^{2+}	0.5 ppm
5	Co^{2+}	0.5 ppm
6	Fe^{2+}	0.1 ppm
7	Cd^{2+}	10 ppm
8	Mn^{2+}	2 ppm

Fig. 7.6g. *Cation separation after PAR derivatization*

Column: Shimadzu IC-C1, 15 cm × 5 mm
Mobile phase: 0.4 mol dm^{-3} lactic acid adjusted to pH 2.85 with NaOH
Flow rate: 1 cm^3 min^{-1}, Temperature = 40 °C
Detector: UV absorption, 520 nm
Reagent: 0.2 mmol dm^{-3} PAR, NH_4OH 3 mol dm^{-3}, CH_3COOH
               1 mol dm^{-3}; 0.3 cm^3 min^{-1} at 40 °C.

7.6.7. Ion Suppression and Ion Pair Chromatography

Many separations previously done by ion exchange are now achieved more easily by the use of ion suppression or ion pairing techniques. Ion suppression is used for the chromatography of weak acids or bases. The principle is that we suppress the ionization of an acid or the protonation of a base by adjusting the pH, and then chromatograph the sample on a reverse phase column (e.g. C-18) using methanol or acetonitrile plus a buffer solution as the mobile phase. The technique is preferable to ion exchange because, compared to ion exchange columns, the C-18 column has higher efficiency, equilibrates faster and is generally better behaved.

Ion pairing techniques are also used to separate weak acids and bases but, additionally, they find application in the separation of other ionic compounds. The methods originated in the field of solvent extraction. An ionized compound (A_{aq}^{+}) that is water soluble can be extracted into an organic solvent by using a suitable counter ion (B_{aq}^-) to form an ion pair, according to the equation:

$$A_{aq}^+ + B_{aq}^- \rightleftharpoons (A^+B^-)_{org}$$

The ion pair (A^+B^-) behaves as if it is a nonionic polar molecule, soluble in organic solvents. By choosing a suitable pairing ion and adjusting its concentration, the ion A^+ can be efficiently extracted into an organic phase. Similarly, anions can be extracted by using a suitable cationic pairing ion. In ion pair chromatography we use a reverse phase separation on a C-18 column, with the ion pairing reagent added to the mobile phase. The ion pairs are separated as neutral polar molecules.

There is still some debate about the mechanism in this method of separation. The simplest model assumes that ion pairs are formed in the mobile phase and travel through the system as neutral species. Separation occurs by partitioning of these neutral ion pairs between the mobile phase and the C-18. This mechanism cannot explain all the experimental results, and there is no doubt that it is a considerable over-simplification. Recent ideas about the mechanism suggest that it involves a combination of partition and ion exchange.

Typical ion pairing reagents are, for cations, alkanesulphonic acids, e.g. pentane, hexane, heptane or octanesulphonoic acid, and for anions, tetrabutylammonium salts. In ion pair chromatography the retention of solutes can be controlled in a number of ways:

(a) By varying the chain length of the pairing agent. Retention increases as chain length increases.

(b) By varying the concentration of the pairing agent. Retention increases as the amount of the pairing agent increases.

(c) If we use a pH at which some solutes in our sample are ionized whilst others are not, then only the ionized solutes will be affected by any changes in the type or concentration of the pairing agent; we can alter the column selectivity for the ionized solutes without affecting the retention of the others.

(d) By changing the concentration of organic solvent in the mobile phase.

∏   How would a change in the concentration of organic solvent affect
      retention?

This will follow the rules for reverse phase separations (Fig. 7.5a).
Increasing the amount of organic solvent will decrease the polarity of the
mobile phase and will decrease retention. Fig. 7.6h shows the separation
of weak acids on a C-18 column with a mobile phase of methanol/water
50 : 50 + tetrabutylammonium phosphate. The pH of the mobile phase is
about 7.5, as at this pH both weak acids are fully ionized and form ion
pairs with the pairing agent.

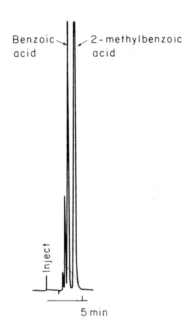

Fig. 7.6h. *Ion pair separation of weak acids*

Column: 10 μm C-18 bonded phase, 30 cm $\times$ 4 mm
Mobile phase: CH_3OH/H_2O 50 : 50 + tetrabutylammonium phosphate, pH
                7.5. Flow rate 2 cm^3 min^{-1}
Detector: UV absorption, 254 nm

Fig. 7.6i(i) shows the use of a combination of ion pairing and ion
suppression to separate a mixture of acids and bases. The pH of the mobile
phase is about 2.5, as at this pH the maleic acid is non-ionized and elutes

quickly as a very polar molecule on the reverse phase column. The other solutes are all weak bases which at pH 2.5 are fully protonated and pair with the pentanesulphonic acid anion.

The retention of the paired solutes can be increased by increasing the chain length of the pairing agent, as shown in Fig. 7.6i(ii).

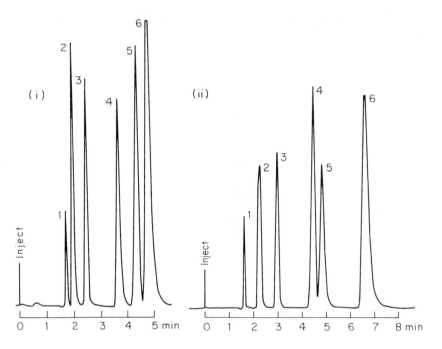

Fig. 7.6i. *Separation of a mixture of acids and bases by ion pairing/ion suppression*

(*i*) Column: 10 μ m C-18 bonded phase, 30 cm × 4 mm
    Mobile phase: CH$_3$OH/H$_2$O 50:50 both containing 0.005 mol dm^{-3} pentane sulphonic acid pH 2.5.
    Flow rate: 2 cm^3 min^{-1}
    Detector: UV absorption, 254 nm
    Sample: 1 maleic acid   2 phenylephrine   3 norephedrine
                      4 naphazoline   5 phenacetin   6 pyrilamine

(*ii*) Conditions as in (*i*) except that hexanesulphonic acid is used as the ion pair reagent.

Ion pairing techniques have also been used in the separation of inorganic ions. Fig. 7.6j shows an example. Detection here is by indirect UV absorption. As the ions themselves do not absorb strongly, an absorbing substance (potassium hydrogen phthalate) is added to the mobile phase. The solute anions, being transparent at the detection wavelength, cause decreases in absorbance, i.e. negative peaks, as they are eluted. Reversing the polarity of the chart recorder enables positive peaks to be recorded.

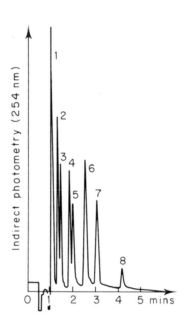

Fig. 7.6j. *Separation of inorganic ions using indirect photometric detection*

Column:	5 μm C-18 bonded phase, 25 cm × 4.6 mm
Mobile phase:	0.001 mol dm^{-3} tetrabutylammonium hydroxide + potassium hydrogen phthalate, pH 8. Flow rate 1 cm^3 min^{-1}
Detector:	UV absorption, 254 nm
Sample:	1 Chloride 2 Nitrite 3 Nitrate 4 Bromide
	5 Phosphate 6 Phosphite 7 Sulphate
	8 Iodide

SAQ 7.6b

Explain the method used in each of the following:

(*i*) Separation of aspirin and norephedrine (1-phenyl-2- aminopropanol)

Column: C-18; mobile phase: CH_3OH/H_2O 50:50 + heptanesulphonic acid (pH about 3.5.).

(*ii*) Chromatography of 4-aminobenzoic acid

Column: C-18; mobile phase: CH_3OH/H_2O 50:50 + tetrabutylammonium hydroxide (pH about 7.5).

7.6.8. Summary

Ionic solutes can be separated by ion exchange chromatography using microparticulate resins or bonded ion exchangers based on microparticulate silica. Such separations are often achieved more easily by ion suppression or ion pairing techniques, which use bonded phase columns in the reverse phase mode.

Learning Objectives

You should now be able to:

• Describe the operation of an ion-exchange material in HPLC;

• Recognize the factors that affect solute retention times in ion exchange chromatography;

• Appreciate that ionic solutes can often be easily separated by ion suppression or ion pairing techniques;

• Specify the experimental conditions that are used in ion pair chromatography.

References

1. F.C. Smith and R.C. Chang, *The Practice of Ion Chromatography*, Wiley, New York, 1982.

2. J.G. Tarter, *Ion Chromatography*, Marcel Dekker, 1987.

3. J.S. Fritz, *Analytical Chemistry* 1987, 59, 35A.

4. T.H. Jupille and D.T. Gjerde, *Journal of Chromatographic Science* 1986, 24, 427.

5. B.D. Bidlingmeyer, *Journal of Chromatographic Science* 1980, 18, 525.

7.7. ADSORPTION AND EXCLUSION CHROMATOGRAPHY

7.7.1. Adsorption Chromatography

Unmodified silicas are used in this mode of HPLC. The adsorption sites on the surface of silica are silanol (Si–OH) groups. These can be present as isolated groups, or can be hydrogen bonded to one another. With chromatographic silicas, the relative numbers of each type of silanol group depend on the type of silica and how it has been prepared and treated. The two types of silanol group have different absorptive strength, and can be deactivated by moisture or by other polar solvents. Usually, for chromatography, the silica is activated by heating at 150–200 °C and then partly deactivated by the addition to the mobile phase of small amounts of water or some other polar organic solvent. This is done to try to standardize the activity of the adsorbent, but reproducibility is always a problem with chromatography on unmodified silica, and to obtain reproducible behaviour often requires lengthy conditioning procedures. With silica and a non-polar mobile phase modified with a small amount of polar solvent, it is possible that a layer of the polar component is adsorbed at the silica surface, and that both adsorption and partition (between this layer and the mobile phase) are contributing to the separation.

The dissolved solute molecules X compete with mobile phase molecules S for a place on the adsorbent surface:

$$X + S_{ads} \rightleftharpoons X_{ads} + S$$

The strength of the interaction between the adsorbent and the solute molecules increases as the polarity of the solute increases. Thus we can increase retention of our solute molecules (X) by decreasing the polarity of the mobile phase (S), which will shift the equilibrium above to the right. Polar solute molecules are strongly held on unmodified silica and tail badly, so the method is useful only for solutes having low or medium polarity.

Fig. 7.7a shows the chromatogram of some phthalates on a silica column using ethyl ethanoate/*iso*-octane 5:95 as the mobile phase. Some of the peaks are identified.

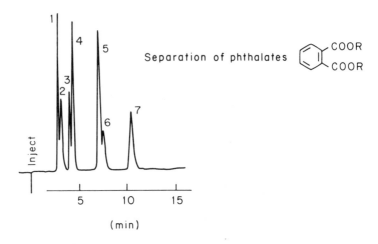

Fig. 7.7a. *Separation of phthalates*

Column: 10 μm silica 30 cm × 4mm
Mobile phase: ethyl ethanoate/*iso*-octane 5:95
Flow rate: 2 cm^3 min^{-1}
Detector: UV absorption, 254 nm
Sample: R = CH$_3$, C$_2$H$_5$, C$_6$H$_5$ (6), *n*-C$_4$H$_9$,
                  iso-C$_4$H$_9$ (3), *iso*-C$_8$H$_{17}$ (2), *n*-C$_8$H$_{17}$

∏   (*a*) In what order would you expect the other solutes to elute?

     (*b*) What changes would you expect in the chromatogram if the mobile phase was changed to butyl ethanoate/*iso*-octane 5:95?

(*a*) In a normal phase separation like this one the solutes elute in order of increasing polarity, so the order of elution would be, first, dioctyl followed by dibutyl, diethyl, dimethyl. You should compare this chromatogram with Fig. 8.3i which involves the same kind of sample but on a reverse phase system. The order of elution is reversed with the most polar solute eluting first.

(*b*) The mobile phase is now a little less polar, so we would expect the retention of all the solutes to increase. The chromatogram is shown in Fig. 7.7b. Changing the mobile phase will also change the selectivity and you can see that this has happened if you look at the diethyl and diphenyl peaks, where the order of elution has been reversed.

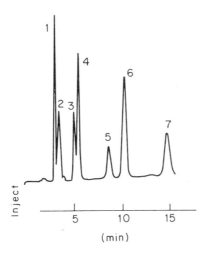

Fig. 7.7b. *Separation pf phthalates*

Column: 10 μm silica 30 cm $\times$ 4 mm
Mobile phase: butyl ethanoate/*iso*-octane 5 : 95
Detector: UV absorption, 254 nm
Sample: 1 R = n-C_8H_{17} 2 *iso*-C_8H_{17} 3 *iso*-C_4H_9 4 n-C_4H_9
                    5 C_6H_5 6 C_2H_5 7 CH_3

7.7.2. Exclusion Chromatography

Exclusion chromatography is a technique for separating molecules based on their effective size and shape in solution. The technique is often called gel permeation chromatography if used with organic solvents or gel filtration if used with aqueous solvents.

The stationary phases used in exclusion chromatography are porous particles with a closely controlled pore size. Unlike other chromatographic modes, in exclusion chromatography there should be no interaction between the solute and the surface of the stationary phase.

Depending on their size and shape, solute molecules may be able to enter the pores of the stationary phase particles. Molecules comparable in size with the mobile phase molecules will be able to diffuse throughout the entire porous network. Larger molecules may be excluded from the narrower parts of the porous structure but will be able to move freely in the wider passages. The larger the solute molecule, the fewer places in the porous structure it will find that it can get into. Finally, there may be solute molecules that are so large that they are completely excluded from the pores. These excluded molecules can travel only through the relatively wide channels between the stationary phase particles, and so are eluted rapidly from the column. The smaller the molecule, the more easily it will be able to penetrate the pore structures of the stationary phase particles, and the longer it will be retained on the column. The process is illustrated in Fig. 7.7c.

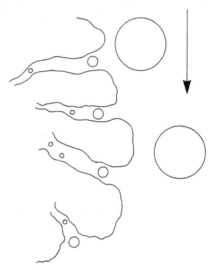

Fig. 7.7c. *Separation by exclusion*
Large molecules are excluded from the internal pores of the stationary phase, and therefore travel in the channels between the stationary phase particles. Smaller molecules that can enter the porous network travel more slowly

In exclusion chromatography, the total volume of mobile phase in the column is the sum of the volume external to the stationary phase particles (the void volume, V_0) and the volume within the pores of the particles (the interstitial volume, V_i). Large molecules that are excluded from the pores must have a retention volume V_0, small molecules that can completely permeate the porous network will have a retention volume of $(V_0 + V_i)$. Molecules of intermediate size that can enter some, but not all, of the pore space will have a retention volume between V_0 and $(V_0 + V_i)$. Provided that exclusion is the only separation mechanism (i.e. no adsorption, partition or ion exchange), the entire sample must elute between these two volume limits.

In Fig. 7.7d the relative molecular mass of the solute, M_r, is plotted on a log scale against the retention volume. The interstitial volume, which represents the volume range within which separations occur, and the size range of solutes that are eluted in this volume range, depend on the sort of material that is used for the stationary phase. Because, for a given separation, V_0 and V_i are constant, we can reliably predict the total volume of solvent or the time taken for a particular analysis. The calibration curve is established by determining the retention volume for standards of known M_r.

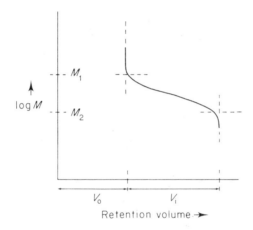

Fig. 7.7d. *Exclusion calibration curve*

Molecules with relative molecular mass $> M_1$ are totally excluded from the stationary phase and have retention volume V_0. Molecules with relative molecular mass $< M_2$ totally permeate the stationary phase and elute at (V_0+V_i). Molecular sizes in between these two partly permeate the stationary phase and elute between V_0 and $(V_0 + V_i)$.

The materials originally used for exclusion chromatography were semirigid gels of cross-linked dextran (a carbohydrate) or polyacrylamide. These cannot withstand the high pressures used in HPLC. Modern stationary phases used in this technique are microparticulate materials consisting of styrene–divinylbenzene copolymers, silica, or porous glass. They are available in a range of pore sizes for the separation of different M_r

ranges. The styrene–divinylbenzene types are used with organic solvents, as aqueous solvents often cause excessive shrinkage of the column bed, producing voids and channels which lead to a loss of efficiency. Fig. 7.7e shows the available sizes and M_r ranges of a styrene–divinylbenzene exclusion stationary phase (Ultrastyragel, manufactured by Waters). It has a particle size of 10 μm and produces efficiencies of around 46 000 plates m^{-1}. Apart from the 100Å type, they can be used at temperatures up to 145 °C.

Pore size, Å	M_r range
100	50–1500
500	100–10^4
10^3	200–3 × 10^4
10^4	5 × 10^3–6 × 10^5
10^5	5 × 10^4–4 × 10^6
10^6	2 × 10^5–10^7

Fig. 7.7e. *Styrene–divinylbenzene exclusion stationary phase (Ultrastyragel, manufactured by Waters)*

The pore size of these is given in Ångstroms (1 Å = 10^{-8} cm) and relates to the chain length of a polystyrene molecule just large enough to be totally excluded from all the pores of the gel. The mass range is the range of M_r (determined using polystyrene standards) that is partially excluded. To provide partial exclusion over a wide range of M_r, a number of columns can be used in series, each column containing a different molecular size range.

∏   A 7.8 mm × 30 cm column containing the 10^4 Å stationary phase above is used with methylbenzene as the mobile phase at a flow rate of 1.1 cm^3 min^{-1}. A sample of polystyrene standards dissolved in benzene is injected. The standards have molecular masses of 775 000, 442 000, 6200 and 2800. The void volume of the column is 6 cm^3 and the interstitial volume is 5 cm^3.

(*a*) Which is the first solute to elute, and what is its retention volume?

(*b*) Which of the solutes are partially excluded?

(*c*) How long does the separation take?

(*a*) The 775K standard will be totally excluded, and will elute at $V_0 = 6$ cm^3;

(*b*) The 442K and the 6.2K standards will be partially excluded;

(*c*) The 2.8K standard and the benzene will totally permeate the column and will elute together at $(V_0+V_i) = 11$ cm^3. This takes $11/1.1 = 10$ min.

Silica stationary phases for exclusion can be used with either organic or aqueous solvents. Some types are bonded phases, others are unmodified. When aqueous phases are used with silica exclusion columns, small amounts of polar mobile phase modifiers (inorganic salts or polar organic solvents) often have to be used to reduce adsorption effects.

The choice of mobile phase for exclusion chromatography is simpler than for other HPLC modes, as only one solvent is required. For polymers, solubility considerations often govern the choice of mobile phase. Tetrahydrofuran or chlorinated hydrocarbons are often used for polymers that are soluble at room temperature. Some polymers (e.g. polyethene) require temperatures of around 150 °C for solution; for these, di- or trichlorobenzenes can be used as the mobile phase.

Although exclusion chromatography was originally used mainly for the characterization of polymers, it now has many uses in other fields. Synthetic polymers have a range of molecular size, and the distribution of M_r will affect many of the important bulk properties of the polymer such as hardness, brittleness, tensile strength and so on. Small changes in M_r distribution may cause large variations in the end-use performance of a polymer. Before the advent of exclusion chromatography, the determination of M_r distribution in polymers was commonly done by fractional precipitation methods, which were difficult, lengthy and inaccurate. These days, the mass distribution and the average M_r of the polymer can easily be calculated from the exclusion chromatogram.

Exclusion chromatography is also useful in the separation of small molecules from interfering matrices of larger molecules, for example in foods or other samples of biological origin. It can be used as the first step in the sequential analysis of complete unknown organic mixtures, which are first separated on a size basis by exclusion, then the collected fractions can be further separated by normal or reverse phase chromatography, where the separation is based on chemical differences.

Fig. 7.7f shows an exclusion chromatogram on unmodified silica. The sample is an epoxy resin with an average M_r of 900. Fig. 7.7g shows the determination of pesticide residues in a sample of chicken fat, and is an example of how exclusion can be used to clean up complex samples. First, a pesticide-free sample of the fat is run as a blank, then the blank is spiked with the pesticides to determine their retention volumes. When the sample is injected, the eluent containing the pesticides is collected. The solvent is evaporated, the residue is dissolved in acetonitrile and the pesticides are then separated on a reverse phase column.

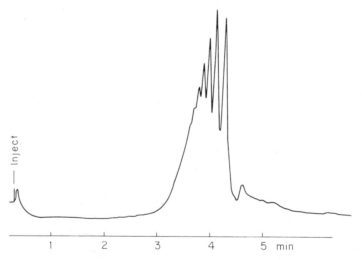

Fig. 7.7f.   *Exclusion chromatogram of an epoxy resin*

Column: 5 μm silica, 25 cm × 4.9 mm
Mobile phase: tetrahydrofuran + 1% H_2O. Flow rate 1 cm³ min⁻¹
Detector: UV absorption, 254 nm
Sample: 1 μl Epikote 1001 in tetrahydrofuran
Epikote 1001 is a synthetic epoxy resin with an average relative molecular mass of 900

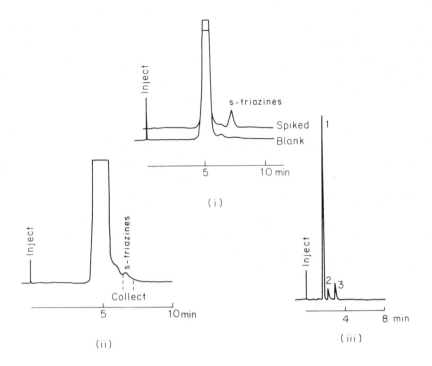

Fig. 7.7g. *Clean up of a sample of chicken fat containing pesticides, using exclusion chromatography*

(*i*) Fat sample, blank and spiked with pesticides
(*ii*) Fat sample containing pesticides

Column: 100 Å μ-Styragel (styrene-divinyl benzene microparticulate resin), 122 cm × 7.8 mm
Mobile phase: trichloromethane, flow rate 2 cm³ min⁻¹
Sample: 100 μl chicken fat in trichloromethane

(*iii*) Reverse phase separation of pesticide residues.

Column: 10 μm C-18, 30 cm × 4 mm
Mobile phase: CH_3CN/H_2O 60:40, flow rate 2 cm³ min⁻¹
Detector: UV absorption, 254 nm.
Peaks: 1 Simazine   2 Atrazine   3 Propazine

SAQ 7.7a

The following statements refer to the different modes of HPLC. Indicate whether the statements are true (T) or false (F).

(*i*) In adsorption chromatography, a non-polar mobile phase is used.

(*ii*) Polar molecules can easily be separated by adsorption chromatography.

(*iii*) The retention times of solutes in exclusion chromatography can be altered by changing the polarity of the mobile phase.

(*iv*) Exclusion chromatography is useful only for the separation of large molecules.

(*v*) In reverse phase chromatography, the mobile phase is more polar than the stationary phase.

(*vi*) In reverse phase chromatography using bonded silica packings, the bonded group is non-polar.

SAQ 7.7b

Which mode of HPLC would you choose for each of the following:

(*i*) Identification of plasticizers in polychloroethene (polyvinyl chloride). Common PVC plasticizers are dibutyl, dioctyl and dinonyl phthalates.

(*ii*) Separation of tranquillizers.

Structures:

Diazepam (valium): $R_1 = -CH_3$, $R_2 = -H$

Oxazepam (serax): $R_1 = -H$, $R_2 = -OH$

(*iii*) Separation of a mixture of synthetic food dyes.

Structures:

Amaranth: $R_1 = R_2 = -SO_3Na$, $R_3 = -H$

Ponceau 4R: $R_1 = -H$, $R_2 = R_3 = -SO_3Na$

SAQ 7.7b

7.7.3. Summary

Adsorption chromatography uses unmodified silica with a relatively nonpolar mobile phase. It is used for the separation of solutes of relatively low polarity, although such separations are now often achieved more easily on bonded phases.

Exclusion chromatography separates solutes that differ in size and shape. The technique is used extensively in the investigation of macromolecules and in the separation of small molecules from an interfering matrix of larger molecules.

Learning Objectives

You should now be able to:

- Predict the order of elution in simple cases using adsorption chromatography on unmodified silica;

- Appreciate the difficulties of obtaining reproducible behaviour in adsorption chromatography;

- Identify stationary and mobile phases used in exclusion chromatography;

- Understand the mechanism by which solutes are separated in exclusion chromatography.

- Recognize the areas in which exclusion chromatography is used.

References

ADSORPTION

1. L.R. Snyder, *Principles of Adsorption Chromatography*, Marcel Dekker, 1968.

2. C.F. Simpson, Ed. *Techniques in Liquid Chromatography*, Wiley, 1984, Chapter 4.

EXCLUSION

3. W.W. Yau, J.J. Kirkland and D.D. Bly, *Modern Size-Exclusion Liquid Chromatography*, Wiley–Interscience, 1979.

4. Chapter 12 of Reference 2, above.

5. J.H. Knox, Ed. *High Performance Liquid Chromatography*, Edinburgh University Press, 1982, Chapter 7.

SAQS AND RESPONSES

SAQ 7.5a

> A test mixture consisting of phenyl methyl ketone, nitrobenzene, benzene and methylbenzene is to be separated on a C-18 column with a mobile phase of CH_3OH/H_2O 60:40. With these conditions, the ketone is eluted first.
>
> (*i*) In what order are the other solutes eluted?
>
> (*ii*) How would you change the composition of the mobile phase so as to increase the retention of the solutes?
>
> (*iii*) How would the retention of the solutes be affected by using a phenyl bonded phase instead of the C-18?
>
> (*iv*) If the C-18 bonded phase contained unreacted silanol groups, how would the retention of the solutes be affected by end-capping the stationary phase?

Response

(*i*) The more polar the solute, the faster it is eluted, so the order would be phenyl methyl ketone, nitrobenzene, benzene, methylbenzene.

(*ii*) We need to increase the polarity of the mobile phase, so we should increase the amount of water, provided that the mixture will still dissolve.

(*iii*) The phenyl bonded phase is slightly more polar than the C-18, and this will cause the solutes to elute faster, though it may not affect all the solutes to the same extent.

(*iv*) End-capping will slightly increase the percentage of carbon in the stationary phase, and will make it less polar. We would expect this to increase retention, and this is what happens for the non-polar solutes benzene and methylbenzene. For the other two, there is another effect to consider, because by end-capping we are reducing the adsorption of the polar solutes on to the stationary phase, and this will decrease their retention. In fact, for the two polar solutes there is a small decrease in retention on end-capping.

****

SAQ 7.6a

For these problems you need to know the pK_a values given below:

	pK_{a1}	pK_{a2}	pK_{a3}
H_3PO_4	2.13	7.20	12.36
H_2CO_3	6.35	10.33	
HF	3.17		
HCl	<1		

(*i*) In what form or forms is phosphate present at pH 1, at pH 2 and at pH 9? Other things being equal, how would the retention of phosphate change with pH on an anion exchange column?

(*ii*) You are separating carbonate and chloride using an anion exchange column and a pH 8.5 buffer as the mobile phase. With this system, carbonate elutes just before chloride. How would you expect the retention of these two to alter if you change the mobile phase to 5×10^{-3} mol dm^{-3} KOH?

(*iii*) Why would it be unwise to run fluoride on an anion exchange column at pH 2.5?

Response

(*i*) At pH 1 predominantly H_3PO_4
At pH 2 H_3PO_4 and $H_2PO_4^-$
At pH 9 predominantly HPO_4^{2-}

As in general the more highly charged ions are more strongly retained, the retention of phosphate would increase as the pH increases.

(*ii*) in 5×10^{-3} mol dm^{-3} KOH, $[OH^-] = 5 \times 10^{-3}$ mol dm^{-3}

$$[H^+] = 10^{-14}/5 \times 10^{-3} = 2 \times 10^{-12} \text{ mol dm}^{-3}$$

$$pH = 11.7$$

$$H_2CO_3 \rightleftharpoons HCO_3^- + H^+ \rightleftharpoons CO_3^{2-} + H^+$$

$$pK_{a1} = 6.35 \qquad pK_{a2} = 10.33$$

At pH 8.5 the carbonate is present as HCO_3^- and at 11.7 as CO_3^{2-}. The doubly charged CO_3^{2-} will be retained more strongly than HCO_3^- and will elute after Cl^-. The retention of the Cl^- would not be affected.

(*iii*) At pH 2.5 an appreciable amount of the fluoride is present as HF. Tailing of the peak would be likely.

****

SAQ 7.6b	Explain the method used in each of the following:
	(*i*) Separation of aspirin and norephedrine (1-phenyl-2- aminopropanol)
	Column: C-18; mobile phase: CH_3OH/H_2O $50:50$ + heptanesulphonic acid (pH about 3.5.). $\longrightarrow$

SAQ 7.6b (cont.)	(*ii*) Chromatography of 4-aminobenzoic acid Column: C-18; mobile phase: CH_3OH/H_2O 50:50 + tetrabutylammonium hydroxide (pH about 7.5).

Response

(*i*) Ion pairing/ion suppression. The aspirin is not ionized in the acid solution whereas the norephedrine, which is a weak base, will be fully protonated and is chromatographed as a neutral ion pair.

(*ii*) Ion pairing. The amine functionality is not protonated at pH 7.5 whereas the carboxylic acid function is fully ionized and pairs with the tetrabutylammonium ion.

****

SAQ 7.7a	The following statements refer to the different modes of HPLC. Indicate whether the statements are true (T) or false (F). (*i*) In adsorption chromatography, a non-polar mobile phase is used. (*ii*) Polar molecules can easily be separated by adsorption chromatography. (*iii*) The retention times of solutes in exclusion chromatography can be altered by changing the polarity of the mobile phase. $\longrightarrow$

SAQ 7.7a   (*iv*) Exclusion chromatography is useful only for the
(cont.)   separation of large molecules.

            (*v*) In reverse phase chromatography, the mobile
               phase is more polar than the stationary phase.

           (*vi*) In reverse phase chromatography using bonded
               silica packings, the bonded group is non-polar.

Response

(*i*) True. The mobile phase should be relatively non-polar.

(*ii*) False. Polar molecules are best separated by reverse phase techniques. Using adsorption, polar molecules have long retention times and suffer from tailing.

(*iii*) False. Provided that exclusion is the only effect that is operating and provided that a change in the mobile phase does not affect the shape of the solute molecule, a change in polarity of the mobile phase will have little effect, as retention in this case is determined by the size and shape of the molecule in relation to the pore size of the stationary phase.

(*iv*) False. Exclusion chromatography is often used to separate small molecules from larger molecules.

(*v*) True.

(*vi*) False. The bonded group is usually non-polar, but the condition for reverse phase operation is given by statement (*v*). Polar bonded groups can be used, provided that the mobile phase is more polar than the stationary phase.

****

SAQ 7.7b	Which mode of HPLC would you choose for each of the following:

(*i*) Identification of plasticizers in polychloroethene (polyvinyl chloride). Common PVC plasticizers are dibutyl, dioctyl and dinonyl phthalates.

(*ii*) Separation of tranquillizers.

Structures:

Diazepam (valium): $R_1 = -CH_3$, $R_2 = -H$

Oxazepam (serax): $R_1 = -H$, $R_2 = -OH$

(*iii*) Separation of a mixture of synthetic food dyes.

Structures:

Amaranth: $R_1 = R_2 = -SO_3Na$, $R_3 = -H$

Ponceau 4R: $R_1 = -H$, $R_2 = R_3 = -SO_3Na$

Response

(*i*) Phthalates can be separated by either normal or reverse phase techniques (see for example Fig. 8.3i) but if you wanted to do this you would either have to extract the plasticizers from the PVC or else risk contaminating your column with high relative molecular mass material. With the correct choice of exclusion column the polymer could be excluded, when it would elute first, followed by the phthalates in order of decreasing relative molecular mass.

(*ii*) The tranquillizers are weak bases, so you have the choice of ion suppression using a C-18 column with a mobile phase of CH_3OH or CH_3CN + alkaline buffer, or ion pairing with a C-18 column and CH_3OH/H_2O + alkane-sulphonic acid.

(*iii*) The strongly ionized sulphonic acid groups preclude ion suppression for these compounds, so for these you can use ion exchange on an anion exchanger, or ion pairing using a C-18 column and CH_3OH/H_2O + a tetrabutylammonium salt as the mobile phase.

****

8. Method Development

When developing an HPLC method, the first step is always to consult the chromatographic literature to find out if anyone else has done the separation, and how they did it. This will at least give you an idea of the conditions that are needed, and may save you having to do a great deal of experimental work. Most textbooks on HPLC have lists of applications, as do a number of journals (e.g. *Analytical Chemistry* annual reviews). At the end of the chapter there is a selection of applications literature that can be obtained from manufacturers.

In this section we will examine some simple case studies to show how a separation is developed or improved. We first take an application quoted by a manufacturer and try and improve it; in the next example we develop a separation with no prior information about the conditions that are needed. Finally, we examine methods for the development of a gradient and for quantitative analysis.

8.1. DETERMINATION OF CAFFEINE IN DECAFFEINATED COFFEE

Fig. 8.1a is taken from the catalogue of a chromatography supplier and shows chromatograms of two coffee samples. The packing used is an irregular 5 μm C-18 modified silica.

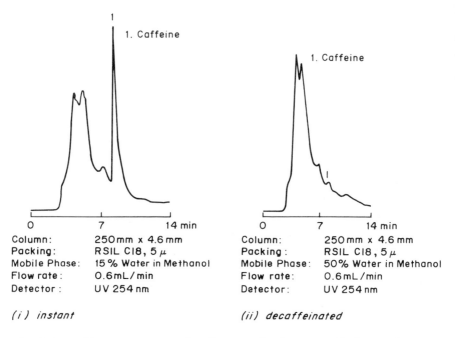

Fig. 8.1a. *Chromatograms of coffee sample, from supplier's catalogue*

∏   What is wrong with the chromatogram in (*ii*) and can you suggest any easy ways to improve the separation? You will find it helpful to look at the UV spectrum of caffeine, Fig. 8.4d.

The chromatogram is truly awful, with the caffeine barely visible on a large tailing band of unresolved peaks. We could make a substantial improvement simply by changing the detection wavelength from 254 to 273 nm, which will give much higher sensitivity. We also want to try and separate the caffeine peak from the rest of the chromatogram, so that the caffeine can be determined quantitatively. You may have noticed as well that the two mobile phase compositions quoted don't make sense, as the caffeine is eluted at the same time in each chromatogram (presumably there was a printing error).

We now run the chromatogram shown in (*ii*) ourselves, trying to use conditions that are roughly similar (apart from the detection wavelength). The sample was prepared by dissolving decaffeinated coffee granules in the mobile phase and filtering before injection. Fig. 8.1b shows the chromatogram obtained.

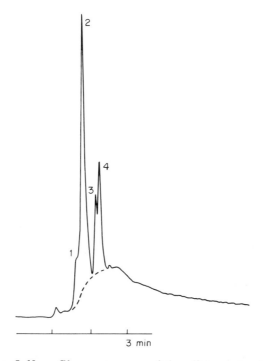

Fig. 8.1b. *Chromatogram of decaffeinated coffee*

Column: Econosphere 5 μm C-18, 15 cm × 4.6 mm
Mobile phase: CH$_3$OH/H$_2$O 50:50
Flow rate: 1 cm^3 min^{-1}; Injection volume: 20 μl
Detector: UV absorption, 273 nm

Π   How could you establish which peak in the chromatogram is due
     to caffeine?

Running a caffeine standard and comparing retention times, or spiking the
sample with a little pure caffeine shows that the caffeine peak is no. 4. As
further confirmation we could run the chromatogram at different detection
wavelengths and look at the absorbance ratios we obtain from peak 4 (see
Section 5.4).

Although this chromatogram is an improvement, it is still not suitable for
quantitative work. Peaks 3 and 4 are not properly resolved and there seems
to be a broad unresolved peak or group of peaks that coelutes (this is drawn
as a dotted line in the figure). To try to resolve peaks 3 and 4 we first alter
the mobile phase composition to increase retention. The capacity factors

are: $k_3' = 1.23$, $k_4' = 1.35$, giving $\alpha_{4,3} = 1.1$; to increase the resolution between 3 and 4 we must first try to increase k'.

∏   How would you alter the mobile phase composition?

In a reverse phase separation we increase retention by increasing mobile phase polarity, so we need to use a mobile phase containing more water. Fig. 8.1c shows the chromatograms obtained using 40% and 30% of methanol respectively. We would expect this change to increase R_S if α remains constant, but in fact there is a favourable change in α as well [in (i) $\alpha_{4,3} \approx 1.3$ and in (ii) $\alpha_{4,3} \approx 1.8$].

The resolution obtained with CH_3OH/H_2O 40:60 is quite adequate, but the caffeine peak is still sitting on a pronounced tail from the earlier part of the chromatogram. To improve this, we use a technique in which the coffee extract is shaken with saturated lead acetate solution and then filtered before chromatography. This removes much of the early eluting material that is causing the tailing. Fig. 8.1d shows the final result. The quantitative analysis of this sample is examined in SAQ 8.4c.

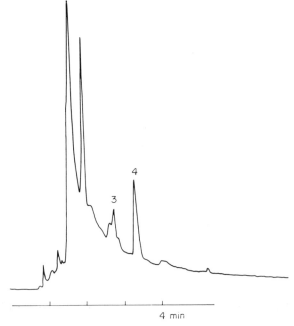

4 min

Fig. 8.1c(i). *Chromatogram of decaffeinated coffee*
Mobile phase: CH_3OH/H_2O 40:60

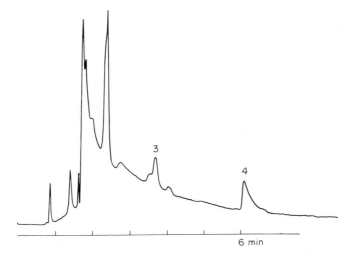

Fig. 8.1c(ii). *Chromatogram of decaffeinated coffee*
Mobile phase: CH_3OH/H_2O 30:70

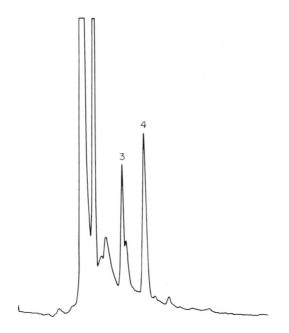

Fig. 8.1d. *Chromatogram of decaffeinated coffee*

Mobile phase: CH_3OH/H_2O 40:60
After lead acetate extraction

LIVERPOOL JOHN MOORES UNIVERSITY
LEARNING SERVICES

8.2. SEPARATION OF STEROIDS

This example shows the steps involved in working out the conditions for the separation of some steroids. Fig. 8.2a shows the structure of some conjugated oestrogens.

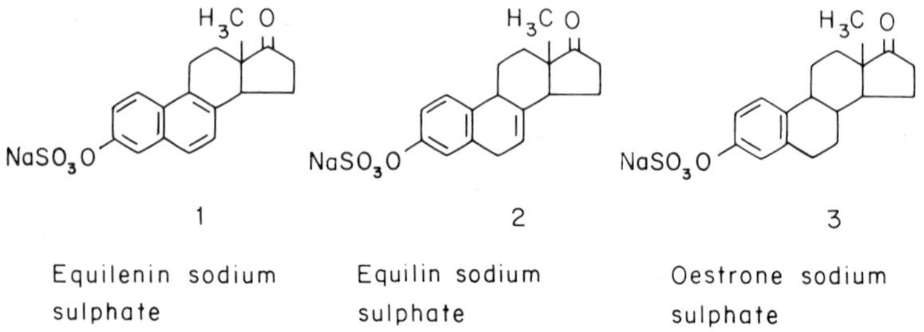

H3C O
NaSO3O
1
Equilenin sodium sulphate
H3C O
NaSO3O
2
Equilin sodium sulphate
H3C O
NaSO3O
3
Oestrone sodium sulphate

Fig. 8.2a. *Structure of conjugated oestrogens*

The task is to separate these from one another, and from excipients, in a commercial tablet. To get an idea of the conditions needed for the separation you have to concentrate on the differences between them.

Π Look at the structures and see if you can suggest:

(*a*) What sort of column packing is needed for the separation;

(*b*) What sort of mobile phase should be used;

(*c*) Whether or not the compounds would be soluble in the mobile phase;

(*d*) Which detector would be the most suitable.

(*a*) If you suggested any sort of non-polar packing then you are thinking along the right lines. The differences between these structures are in the non-polar parts of the molecules, so we need a non-polar packing to exploit these differences, ideally a packing very similar to the parts of the molecules that differ. A phenyl bonded phase would probably be the best bet, but in this case a non-polar C-18 column was used.

(*b*) and (*c*) A C-18 column needs a polar mobile phase such as methanol/water or acetonitrile/water, so as a starting point a 50 : 50 methanol/water mixture was chosen. Because the three compounds are sodium salts, they should be soluble in this mixture. This is easily checked using standards of the three compounds.

(*d*) The aromatic rings suggest that UV absorption would be a suitable method of detection.

Sample preparation consists of crushing some tablets, mixing with 50:50 methanol/water, diluting to the mark in a volumetric flask and then filtering off any insoluble excipients. We are now ready to go, and Fig. 8.2b(i) shows the results of the first injection.

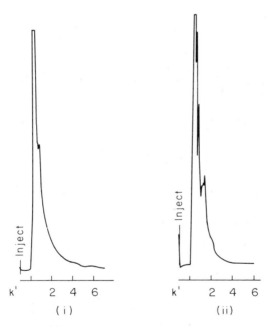

Fig. 8.2b. *Chromatograms of a steroid tablet*

Mobile phase: (*i*) CH_3OH/H_2O 50 : 50
(*ii*) CH_3OH/H_2O 35 : 65
Column: 10 μm C-18 bonded phase, 30 cm $\times$ 4 mm
Flow rate: 2 cm^3 min^{-1}
Detector: UV absorption, 254 nm

This is a disaster, as there is little or no separation, and almost everything is eluting from the column immediately, at k' a little more than zero. The first step is to increase the retention times of all the solutes, i.e. increase the k' values.

∏   Would you do this by:

   (a) Increasing the amount of methanol in the mobile phase;

   (b) increasing the amount of water in the mobile phase;

   (c) changing to an acetonitrile/water mobile phase?

As in the last example, the way to increase k' in a reverse phase separation is to increase the polarity of the mobile phase, so we need to increase the amount of water. It would not be sensible to change to acetonitrile/water at this stage.

In Fig. 8.2b(ii) the amount of water in the mobile phase has been increased to 65%. This chromatogram is not much better than the first one, but we are starting to get longer retention times and some resolution. We need to know if the things that are starting to separate out are the oestrogens or just rubbish from the excipients in the tablet. The next two chromatograms (Fig. 8.2c) are for mixed standards using mobile phases containing, respectively, 65% and 80% water. This shows that the oestrogens are indeed being retained a little longer than the excipients, so that we are going in the right direction.

The next chromatogram, Fig. 8.2d(i), is an injection of the tablet solution with the 80% water mobile phase. You can see from this that the oestrogens have k' values between 3 and 5 and that they are separated from the excipients. Although things are improving, there are still several problems left to solve.

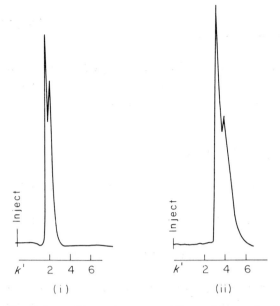

Fig. 8.2c. *Chromatograms of steroid standards*

Mobile phase: (*i*) CH$_3$OH/H$_2$O 35 : 65
(*ii*) CH$_3$OH/H$_2$O 20 : 80

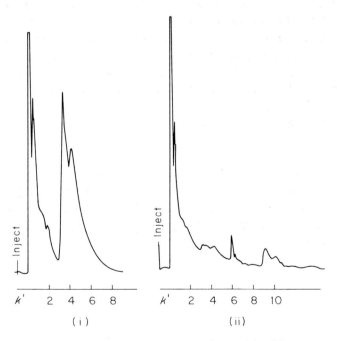

Fig. 8.2d. *Chromatograms of steroid tablet*

Mobile phase: (*i*) CH$_3$OH/H$_2$O 20 : 80
(*ii*) CH$_3$OH/0.001 mol dm^{-3} KH$_2$PO$_4$ 20 : 80 (pH 5)

∏   Can you see two undesirable features in the chromatogram?

These are: (a) the oestrogens are not properly resolved (there are only two peaks) and (b) the oestrogen peaks are tailing.

The tailing is probably caused by a mixed mechanism, for instance, adsorption on active silica sites that are not end-capped. To reduce this, we can try adding a salt to the water. To get better resolution we need to change the selectivity, α , which means changing the chemistry of the mobile phase, or increasing the plate number of the column, or both.

∏   Do you think that the addition of a salt as a modifier will affect k', α or N (or a combination of them) and, if so, what will happen?

Addition of a salt will have no effect on N but will have a drastic effect on the k' values, as we are increasing the polarity of the mobile phase. We would expect k' values to increase.

Because the addition of a salt will change k' and probably the α values, we need to try this first. Fig. 8.2d(ii) shows the chromatogram obtained using $0.001 \text{ mol dm}^{-3}$ KH_2PO_4 (pH 5) instead of the water. What has happened here is that the oestrogens either do not elute at all or, if they do, elution takes far too long. This is not surprising, because we have made the mobile phase a great deal more polar and forced the oestrogens to interact more strongly with the C-18.

It is tempting at this point to say that we made a mistake in adding the phosphate, but in fact there is no way of knowing the effect the phosphate has had on tailing and selectivity until we can get the k' values back into the 1–10 region and have a look at the peaks. So we now need to decrease the k' values, which means changing the methanol/water ratio yet again. Fig. 8.2e shows the chromatogram obtained using phosphate, but with the original 50 : 50 methanol/water ratio. The k' values are in the required range, the tailing problem has been eliminated, and the selectivity is better.

Fig. 8.2e. *Chromatogram of steroid tablet*

Mobile phase: $CH_3OH/0.001$ mol dm^{-3} KH_2PO_4 50:50

The resolution of the three oestrogens still has to be improved, so to proceed further we can either work on the selectivity by using a different water-soluble solvent or a mixture, or we can try to improve the separation by increasing the plate number of the column. The best solution would probably be to try to optimize a mixed solvent composition, as described in Chapter 6. The separation can be improved, at the expense of a longer analysis time, by using two columns in series, as shown in Fig. 8.2f.

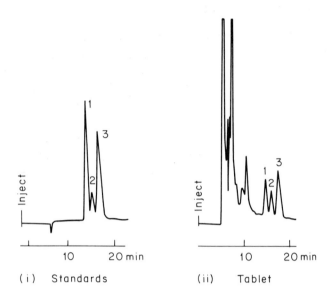

Fig. 8.2f. *Chromatograms of steroids*

Column: Two 30 cm × 4 mm columns containing 10 μm C-18 bonded phase
Flow rate: 1 cm^3 min^{-1}
Mobile phase: CH$_3$OH/0.001 mol dm^{-3} KH$_2$PO$_4$ 50:50
Detector: UV absorption, 254 nm

8.2.1. Summary

Experimental conditions are developed for the separation of caffeine in instant coffee and for steroids in a commercial tablet.

Learning Objectives

You should now be able to:

- suggest a suitable column, mobile phase and detector that might be used for a given separation;

- suggest how the column and/or the mobile phase composition might be altered so as to improve resolution and peak shape.

8.3. GRADIENT ELUTION

In a sample containing many different solutes, with isocratic elution, it is sometimes impossible to choose a suitable mobile phase that will result in all k' values being within the optimum range. If this is the case, the chromatogram may appear as in Fig. 8.3a. The early peaks appear at k' values between 0 and 1 and are poorly resolved. Peaks 5 and 6 are well resolved but peaks 7 and subsequent peaks are getting very dispersed and are taking a long time to elute. To improve the chromatogram we need to increase the k' values of the early peaks by using a weaker mobile phase and to decrease the k' values of later peaks by using a stronger mobile phase.

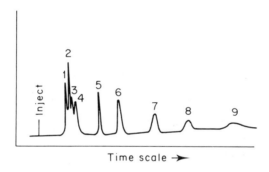

Time scale →

Fig. 8.3a. *Chromatogram under isocratic conditions*

The problem can be solved by the use of gradient elution, where the composition of the mobile phase is altered during the separation. This can be done in a number of ways, as shown in Fig. 8.3b.

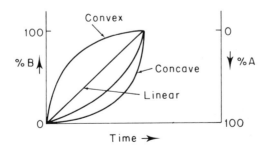

Fig. 8.3b. *Possible gradient profiles showing the blending of solvent B with solvent A*

Modern solvent delivery systems allow you to select a wide range of gradient profiles and to vary the time over which the gradient is delivered. As well as the shape and time of the gradient, we have to consider a number of other factors such as compatibility of the two solvents with the detector, or miscibility with the sample solution. For instance, if a UV absorbance detector is used and the absorbance of the two gradient solvents is slightly different, then the baseline will drift during the gradient. This can be overcome by adding an unretained absorbing substance to one of the solvents to adjust the solvents to a constant absorbance.

It is important to run a blank gradient and to regenerate the column after the gradient, either by running the gradient in reverse or, for bonded phases, by pumping through about five column volumes of the starting solvent. Fig. 8.3c shows a linear blank gradient run on a C-18 column. The peaks in the chromatogram are artefacts from the distilled water. What happens is that the non-polar column initially concentrates traces of organics from the water at the head of the column. As the solvent polarity is decreased, the organic material is eluted. Clearly, in this case the water needs to be

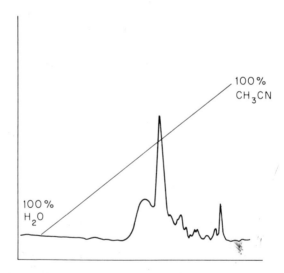

Fig. 8.3c.   *Blank gradient (run without injection of sample)*

Column: 10 μm C-18, 30 cm × 4 mm
Mobile phase: A = H_2O, B = CH_3CN; 0% B → 100% B
Flow rate: 0.5 cm^3 min^{-1}
Detector: UV absorption, 254 nm

purified, but if a blank gradient had not been run we would not have known if the peaks came from the sample or the mobile phase.

8.3.1. Development of a Gradient

This section describes the practical development of a gradient using as a sample Triton X-100, which is a non-ionic surfactant. Chemically, it is a polyethene glycol with a range of relative molecular masses (the average M_r is 624). We want to resolve the different molecular sizes with a normal phase separation on a silica column. Fig. 8.3d shows the chromatogram obtained isocratically with a mobile phase consisting of trichloromethane plus a small amount of dimethyl sulphoxide (DMSO).

Fig. 8.3d. *Chromatogram of Triton X-100. Isocratic elution*

Column: 10 μm silica, 30 cm × 4 mm
Mobile phase: trichloromethane/DMSO 97 : 3
Flow rate: 3 cm³ min⁻¹
Detector: UV absorption, 280 nm

As in the previous example, we have little or no separation, and everything is eluting much too quickly.

∏   To improve matters, would you change the mobile phase to:

(*a*) dioxane/DMSO;

(*b*) dichloromethane/DMSO;

(*c*) heptane/trichloromethane/DMSO;

(*d*) heptane/DMSO?

To increase retention in a normal phase separation we need a less polar mobile phase, so option (*a*) would make things worse. All the other mixtures are less polar than the starting mobile phase, but mixture (*b*) only slightly less so, which probably would not make much difference. Heptane is non-polar but the highly polar DMSO is not soluble in it. It is best to keep the trichloromethane/DMSO and add a modifier, as in (*c*). We can then change the polarity as we wish by altering the relative amounts of trichloromethane and heptane in the mixture.

The next three chromatograms, Fig. 8.3e, show the effect of increasing the amount of heptane in the mobile phase. The mobile phases used were:

(*i*) DMSO/trichloromethane/heptane 3 : 40 : 60

(*ii*) DMSO/trichloromethane/heptane 3 : 20 : 80

(*iii*) DMSO/trichloromethane/heptane 3 : 10 : 87

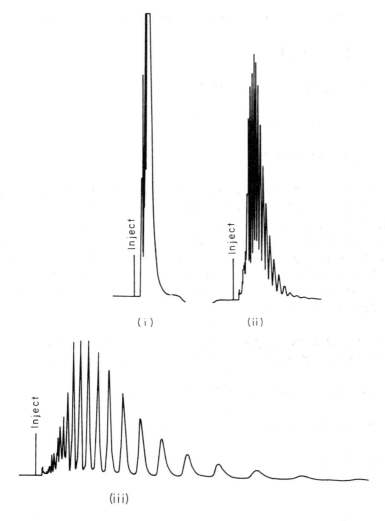

Fig. 8.3e. *Chromatograms of Triton X-100. Isocratic elution*

Π   In Fig. 8.3e(iii), how would you improve the separation using a gradient?

The separation takes rather a long time compared with the others and the later eluting peaks are highly dispersed.

We need a gradient using a more polar solvent in the later stages of the chromatogram. In Fig. 8.3f we use a two solvent gradient consisting of:

Solvent A: DMSO/trichloromethane/heptane 3 : 10 : 87

Solvent B: DMSO/trichloromethane 3 : 97

The starting mobile phase is 100% A and the linear gradient is run over 20 min to a final proportion of 50% B.

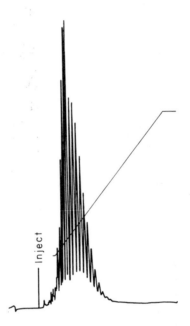

Fig. 8.3f.   *Chromatogram of Triton X-100. Gradient elution*

Π   To improve the resolution would you

(*a*) increase the time over which the gradient is delivered;

(*b*) use a weaker solvent for B (e.g. another mixture of DMSO, trichloromethane and heptane);

(*c*) keep the same solvents but finish up with a smaller proportion of B;

(*d*) use a concave gradient, so that B is added more slowly?

Any of these should produce some improvement. Option (*a*) seems the least promising, as only a small amount of the gradient is actually being used. Of the remainder, choices (*c*) and (*d*) are easier experimentally, as we can do these by reprogramming the gradient former, whereas if we change the mobile phase we have to spend time flushing the system through.

Fig. 8.3g and Fig. 8.3h show the improvement produced by using a combination of these options. In the final chromatogram using a smaller proportion of B added as a concave gradient over a longer time the peaks are now eluted over the whole range of the gradient, and most of them are resolved almost to the baseline.

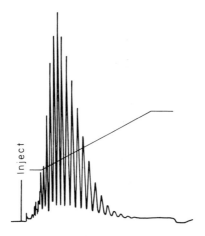

Fig. 8.3g. *Chromatogram of Triton X-100. Gradient elution*

Solvent A: DMSO/trichloromethane/heptane 3 : 10 : 87
Solvent B: DMSO/trichloromethane 3 : 97
Gradient: 0–20% B, 20 min

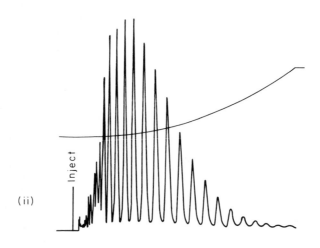

Fig. 8.3h. *Chromatogram of Triton X-100. Gradient elution*

Solvent A: DMSO/trichloromethane/heptane 3 : 10 : 87
Solvent B: DMSO/trichloromethane 3 : 97
Exponential gradient: 0–20% B, 30 min

SAQ 8.3a   Suggest a gradient that would improve each of the chromatograms in Figs 8.3i and 8.3j. You do not need to worry about the detail, like the exact shape of the gradient, or how long it will take. Concentrate on the composition of mobile phase that is needed at the start and at the end of each chromatogram.

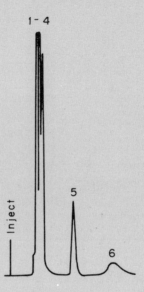

Sample : Six phthalate plasticizers

Fig. 8.3i. *Chromatogram requiring gradient*

Sample: Six phthalate plasticizers
           1 R = CH_3 2 CH_2H_5 3 C_6H_5 4 n-C_4H_9 5 n-C_8H_{17} 6 iso-$C_{10}H_{21}$
Column: 10 μm C-18 bonded phase, 30 cm × 4 mm
Mobile phase: methanol/water 90 : 10
           Flow rate 2 cm^3 min^{-1}
Detector: UV absorption, 254 nm

$\longrightarrow$

SAQ 8.3a
(cont.)

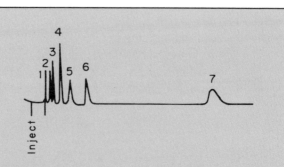

Fig. 8.3j. *Chromatogram requiring gradient*

Sample: 1 benzene   2 diphenyl ether
               3 ethyl benzoate   4 carbazole
               5 nitrobenzene   6 diphenyl ketone
               7 benzyl alcohol
Column: 5 μm silica, 30 cm × 1.8 mm
Mobile phase: dichloromethane/hexane 40:60
               Flow rate 2 cm^3 min^{-1}
Detector: UV absorption, 254 nm

8.3.2. Summary

With isocratic elution and a sample having solutes with a wide range of polarity it is sometimes not possible to achieve the desired resolution in an acceptably short time. It may be possible to improve the chromatogram using gradient elution. A practical example of the development of a gradient is discussed.

Learning Objectives

You should now be able to:

- Recognize a chromatogram where the separation would be improved by the use of a gradient;

- Describe a suitable gradient for a normal and a reverse phase separation;

- Appreciate some of the disadvantages of using gradients.

8.4. QUANTITATIVE ANALYSIS

8.4.1. Area/Height Percent (Also Called Internal Normalization)

For quantitative analysis we assume that the area (or the height) of our peak in the chromatogram is proportional to the amount of substance that produced the peak. In the simplest method we measure areas or heights, which are then normalized (this means that each area or height is expressed as a percentage of the total). The normalized heights or areas give the composition of our mixture, as in the following example:

Peak number	Peak height (mm)	Normalized peak height = % w/w
1	12	$12/162 \times 100 =$ 7.4
2	27	16.7
3	72	44.4
4	_51_	_31.5_
	162	100.0

There are two problems with this approach, which are:

(*a*) We have to be sure that we have counted all the components, i.e. that each component appears as a separate peak on the chromatogram. Components may coelute, or be retained on the column, or may elute without being detected.

(*b*) We are assuming that we get the same detector response for equal amounts of each component. This is seldom the case.

Because of these difficulties, calibration of the detector is usually required. The way that this is done is dealt with in the following sections.

8.4.2.   External Standard Method

In this method we prepare a standard containing the compound or compounds that we wish to determine, ideally present at about the same level as in the unknown, and we compare the chromatograms of standard and unknown. From the standard chromatogram, a response factor is calculated for each peak of interest, the response factor giving us the concentration of component that produces unit detector response:

$$\text{Response factor} = \frac{\text{concentration of component}}{\text{peak height or area}}$$

Then for the unknown chromatogram we can calculate the concentration of each component of interest by multiplying the peak height or area by the appropriate response factor.

For this method to work, the detector response must be linear for each compound over the range of concentration being used, and also we have to inject exactly the same amount of material for each of the two chromatograms, so successful operation of the method depends on our being able to inject samples with good precision.

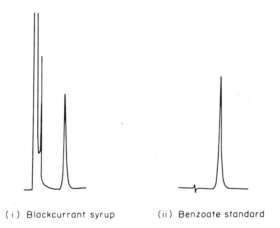

(i) Blackcurrant syrup   (ii) Benzoate standard

Fig. 8.4a. *Determination of benzoate in blackcurrant syrup*

Column: Zorbax 5 μm C-18, 25 cm × 4.6 mm
Mobile phase: CH_3CN/0.005 mol dm^{-3} pH 4.5 acetate buffer 15 : 85
Flow rate: 1.5 cm^3 min^{-1}; temperature: 40 °C
Detector: UV absorption, 254 nm

Fig. 8.4a shows some results obtained for the determination of benzoate added as a preservative to blackcurrant syrup. Chromatogram (*ii*) is a sodium benzoate standard (concentration 0.07308 g dm^{-3} made up in mobile phase), chromatogram (*i*) is blackcurrant syrup (concentration 90.6726 g dm^{-3} in mobile phase). Both chromatograms were run at the same detector sensitivity. Peak areas were measured using an integrator, which prints out a number that is proportional to the peak area. The benzoate peak in the unknown is the peak having the same retenticn time as the benzoate in the standard. Peak areas obtained for the benzoate were:

standard 103 741   unknown 72 859

∏   Calculate the percentage by weight of benzoate preservative in the blackcurrant syrup.

$$\text{Response factor} = \frac{73.08 \text{ ppm}}{103\ 741} = 7.044 \times 10^{-4}$$

Benzoate concentration = 72 859 × 7.044 × 10^{-4} = 51.32 ppm

The blackcurrant syrup was diluted before analysis, so the concentration of benzoate in the original syrup was:

$$51.32 \times \frac{1000}{90.6726} = 566 \text{ ppm or } 0.0566\%.$$

SAQ 8.4a

Fig. 8.4b is for the determination of benzoic acid and methyl and propyl parabens in sennoside. Sennoside is a laxative syrup and the three additives are used as stabilizers and preservatives. The parabens are alkyl esters of 4-hydroxybenzoic acid.

Chromatogram (i) is for a standard mixture of benzoic acid (0.06596 g dm^{-3}), methyl paraben (0.05802 g dm^{-3}) and propyl paraben (0.07470 g dm^{-3}) made up in the mobile phase.

Chromatogram (ii) is sennoside (256.4480 g dm^{-3}) made up in the mobile phase.

(a) What is the order of elution of the three solutes in Fig. 8.4b(i)?

(b) In Fig. 8.4b(ii), identify the peak for each additive in the mixture.

(c) From Fig. 8.4b(ii), using peak height measurement, calculate response factors for each additive and use these to calculate the concentration of each additive in the sample of sennoside.

SAQ 8.4a
(cont.)

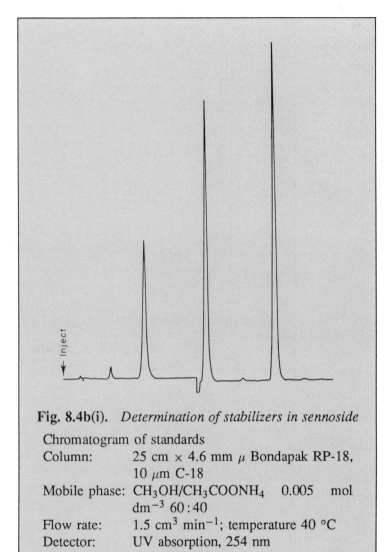

Fig. 8.4b(i). *Determination of stabilizers in sennoside*
Chromatogram of standards
Column: 25 cm × 4.6 mm μ Bondapak RP-18,
               10 μm C-18
Mobile phase: CH_3OH/CH_3COONH_4 0.005   mol
               dm^{-3} 60:40
Flow rate: 1.5 cm^3 min^{-1}; temperature 40 °C
Detector: UV absorption, 254 nm

$\longrightarrow$

SAQ 8.4a
(cont.)

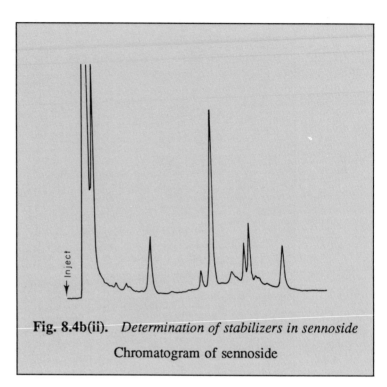

Fig. 8.4b(ii). *Determination of stabilizers in sennoside*
Chromatogram of sennoside

SAQ 8.4a

8.4.3. Internal Standard Method

In this method we add to our sample a known amount of a standard substance (the internal standard). The chromatogram obtained is compared with the chromatogram of a known mixture of the compounds of interest (again containing a known amount of internal standard). This method has all the advantages of the external standard method but, in addition, it compensates for variations in injection volume and also for small changes in detector sensitivity or chromatographic changes that might occur. Because we do not need to inject exactly the same amount each time, this method generally has better precision than the use of an external standard.

From the standard chromatogram we work out response factors for each component of interest, but this time we express the response factor relative to the response factor of the internal standard.

Suppose that, in our standard mixture:

c = concentration of the component of interest

A = peak area (or height) for this component

c_s = concentration of internal standard

A_s = peak area (height) of internal standard

then the relative response

$$r = \frac{c/A}{c_s/A_s} \tag{8.4a}$$

In the unknown mixture let:

c_u = concentration of component

A_u = peak area (height)

c_s' = concentration of internal standard

A_s' = peak area (height) of internal standard

then
$$c_u = A_u \times r \times \frac{c_s'}{A_s'}$$
(8.4b)

Another approach is to correct each peak area in the unknown mixture by multiplying by the appropriate relative response factor. This produces the peak areas that would have been obtained had the detector response been the same for each component. The composition of the mixture is then obtained by normalizing the corrected areas. For this to work we again have to be sure that we have seen each component in the mixture as a separate peak on the chromatogram.

As an example of this method, we will work out response factors for aspirin (acetylsalicylic acid) and caffeine relative to the internal standard phenacetin. Analgesic tablets often contain aspirin and caffeine, and we will eventually use the results for the quantitative analysis of a commercial tablet. The structures of each of these substances are shown in Fig. 8.4c.

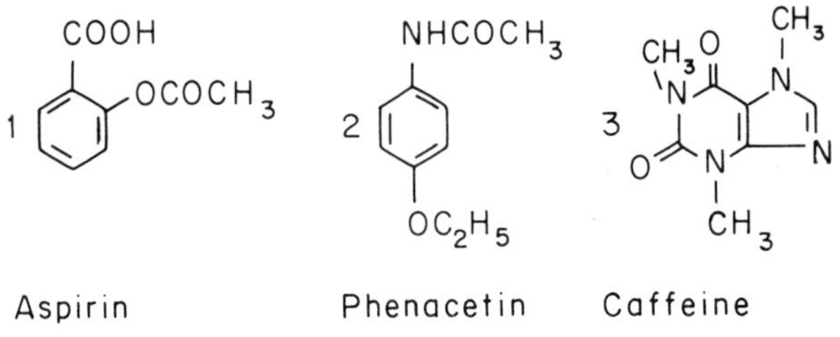

Fig. 8.4c. *Structures of aspirin, phenacetin and caffeine*

To separate the three substances, we will use a method taken from the literature (G.B. Cox et al., *Journal of Chromatography* 1976, 117, 269–278). The column is 12.5 cm × 4.6 mm with a 5 μm silica SCX bonded phase. The mobile phase is 0.05 mol dm^{-3} HCOONH$_4$ + 10% ethanol, pH 4.8, pumped at 2 cm^3 min^{-1} with an inlet pressure of about 117 bar. With these conditions the compounds are separated in about 3 minutes (see Fig. 8.4e).

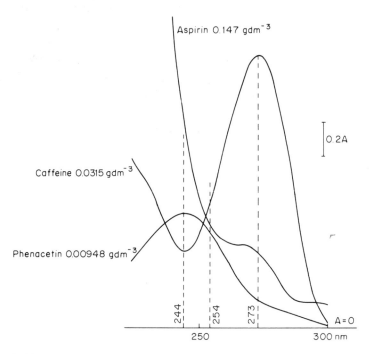

Fig. 8.4d. *UV absorption spectra of aspirin, phenacetin and caffeine*

Fig. 8.4d shows the UV absorption spectra (300–225 nm) of the three compounds, made up in the mobile phase.

∏   Which of the three wavelengths 273, 254 or 244 nm would you choose for the detection of these compounds?

I think 254 nm is the least attractive of the three. At this point the absorbance of each compound is changing rapidly with wavelength and we want, if possible, to monitor the absorbance in a relatively flat region of the spectrum (see Section 5.3). Detection at 273 nm would give a high sensitivity for caffeine, but it might represent rather a low sensitivity for phenacetin (although phenacetin absorbs much more strongly than the other two). I would choose 244 nm, which corresponds to a minimum and a maximum respectively in the spectra of caffeine and phenacetin, although I don't think there is much to choose between 244 and 273 nm.

To work out relative responses, a standard mixture was made up as follows: 0.6015 g of aspirin + 0.0765 g of phenacetin + 0.0924 g of caffeine were dissolved in 10 cm^3 of absolute ethanol. 10 cm^3 of 0.5 mol dm^{-3} ammonium formate was added and the solution was made up to 100 cm^3. Figure 8.4e shows the chromatogram obtained from a 1 μl injection of this solution. Three injections were made, the results of which are shown in Fig. 8.4f.

Fig. 8.4e. *Separation of aspirin, phenacetin and caffeine*

Column: 5 μm silica SCX, 12.5 cm × 4.6 mm
Mobile phase: 0.05 mol dm^{-3} HCOOONH$_4$ + 10% ethanol, pH 4.8
Flow rate: 2 cm^3 min^{-1}
Detector: UV absorption, 244 nm
Peaks: 1 aspirin   2 phenacetin   3 caffeine

∏   Use Eq. 8.4a to calculate response factors for aspirin and caffeine relative to phenacetin = 1. It is not necessary to work out the concentration of each component, you can simply use the mass taken, which is proportional to concentration. Average the response factors for each compound and put them in the figure below. The correct results for this figure are given at the end of the section.

Injection no.	Aspirin	Phenacetin	Caffeine
	mass in mixture, g	0.6015	0.0765	0.0924
1	peak area relative response (*r*)	144 090	159 516 1	43 057
2	peak area *r*	143 200	163 164 1	43 099
3	peak area *r*	121 297	139 796 1	36 564
	Average *r*		1	

Fig. 8.4f.   *To be completed*

These relative response values are used in the SAQ which follows.

SAQ 8.4b   A commercial analgesic tablet is stated on the packet
            to contain 325 mg of aspirin and 50 mg of caffeine
            per tablet. Two tablets plus 0.0773 g of phenacetin
            were shaken with 10 cm^3 of ethanol for 10 min, then
            10 cm^3 of 0.5 mol dm^{-3} ammonium formate was
            added and the mixture was made up to 100 cm^3. The
            tablets contain insoluble excipients, so a little of the
            solution was filtered before chromatography. Fig. 8.4g
            gives peak areas obtained for two 1 μl injections of
            the solution. Chromatographic conditions were those
            described previously.

Injection no.	Peak area		
	Aspirin	Phenacetin	Caffeine
1	157595	170804	50693
2	153541	164174	48478

Fig. 8.4g.

Using Eq. 8.4b and the response factors from the last
section, calculate the amount of aspirin and caffeine
present, expressing the results as mg per tablet.

The aspirin content should be between 95 and 105%
and the caffeine content between 90 and 110% of the
amount stated on the packet. Are the tablets within
specification?

SAQ 8.4b

SAQ 8.4c

The following is an outline of a method for the determination of caffeine in decaffeinated instant coffee.

About 0.8 g of granules is accurately weighed, dissolved and made up to 50 cm³. 5 cm³ of this solution is shaken with 5 cm³ of saturated lead acetate solution for 5 min. The resulting solution is then filtered before injection. A standard solution of caffeine, about 50 ppm, is also injected.

A typical chromatogram is shown in Fig. 8.1d.

(*a*) The sample solution injected should be dissolved in the mobile phase. How would you dissolve the granules to ensure this?

(*b*) Results from a typical determination are shown below:

mass of coffee taken	0.8277 g
caffeine standard: concentration	59.2 ppm
peak area	33 612
coffee solution: caffeine peak area	7262

From these results, calculate the percentage of caffeine in the coffee.

(*c*) A series of 10 determinations on the same sample of decaffeinated instant coffee gave a mean result of 0.149 ± 0.0046% for the caffeine level. Can you suggest any alterations to the experimental procedure that would improve the accuracy and the precision of the method?

SAQ 8.4c

SAQ 8.4d   Draw up a table which lists briefly the advantages and limitations of each of the methods of quantitative analysis that were discussed in the previous section.

SAQ 8.4e

Fig. 8.4h shows the UV spectra of four food additives. Tartrazine is a yellow colouring, saccharin and aspartame are sweeteners. Fig. 8.4i shows the chromatogram of a mixture of these additives, obtained using the following conditions:

Column: 5 μm C-18, 25 cm × 4.6 mm
Mobile phase: CH_3OH/NaH_2PO_4 pH 4.5 using a gradient from 5:95 to 95:5
Flow rate: 1.5 cm^3 min^{-1}
Detector: UV (PDA) using wavelength programme

(*i*) Given that peak 3 on the chromatogram is caffeine, suggest the order of elution of the other solutes.

(*ii*) Suggest, with reasons, the probable outcome if the separation was done using isocratic elution with a mobile phase composition of, say, CH_3OH/NaH_2PO_4 20:80 or CH_3OH/NaH_2PO_4 80:20.

(*iii*) What problems would arise if a single wavelength of 254 or 280 nm was used for detection?

(*iv*) Suggest a suitable wavelength programme that could be used for the separation.

$\longrightarrow$

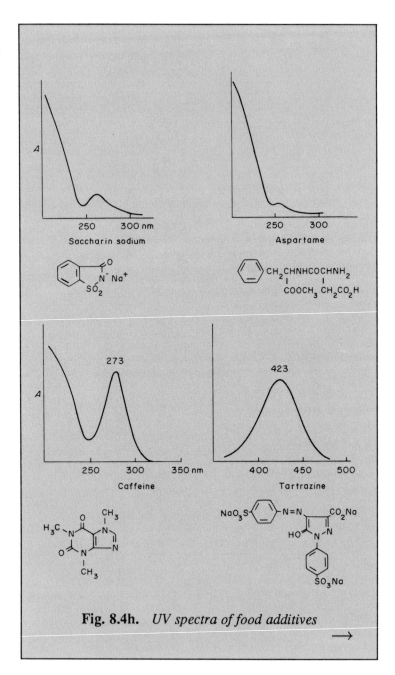

Fig. 8.4h. *UV spectra of food additives*

$\longrightarrow$

SAQ 8.4e
(cont.)

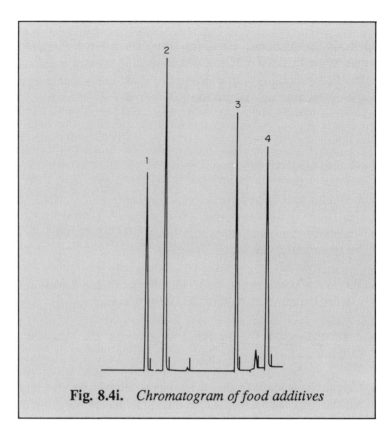

Fig. 8.4i. *Chromatogram of food additives*

8.4.4. Summary

Methods for quantitative analysis using an external standard or an internal standard are discussed. Using UV absorption spectra, a suitable wavelength is found for simultaneous detection of three components in a mixture, and response factors are calculated.

Learning Objectives

You should now be able to

• Recognize the advantages and limitations of the various methods used for quantitative analysis;

• Use UV absorption spectra to identify a suitable detection wavelength to use for the quantitative analysis of a mixture;

• Appreciate that for quantitative analysis the detector needs to be calibrated;

• Calculate response factors from data obtained with standard mixtures and use the response factors for the quantitative determination of an unknown.

Applications Data from Manufacturers

Almost every manufacturer or supplier of HPLC equipment publishes applications journals or notes; the following is a short selection of useful material.

(*a*) *Alltech Chromatography Newsletter*, Alltech Associates Applied Science Ltd., Kellet Road Industrial Estate, Carnforth, Lancs., LA5 9XP.

(*b*) *Chrompack News*, Chrompack UK Ltd., Unit 4, Indescon Court, Millharbour, London E14 9TN.

(*c*) *Column, Life Sciences Notes, Environmental Notes and others*, Millipore UK Ltd., Waters Chromatography Division, The Boulevard, Ascot Rd., Croxley Green, Herts., WD1 8YW.

(*d*) *Analytical News*, Perkin Elmer Ltd., Post Office Lane, Beaconsfield, Bucks., HP9 1QA.

(*e*) *Unicam Analytical Services*, York St., Cambridge, CB1 2PX.

(*f*) *Shimadzu Applications Data Book*, Dyson Instruments Ltd., Hetton Lyons Industrial Estate, Houghton-le-Spring, Tyne and Wear, DH5 0RH.

(*g*) *Supelco Reporter*, Supelco Chromatography Supplies (R.B. Bradley Ltd.), Shire Hill, Saffron Walden, Essex, CB11 3AZ.

(*h*) *Varian LC at Work*, Varian UK Ltd., 28 Manor Road, Walton on Thames, Surrey, KT12 2QF.

SAQS AND RESPONSES

SAQ 8.3a Suggest a gradient that would improve each of the chromatograms in Figs 8.3i and 8.3j. You do not need to worry about the detail, like the exact shape of the gradient, or how long it will take. Concentrate on the composition of mobile phase that is needed at the start and at the end of each chromatogram.

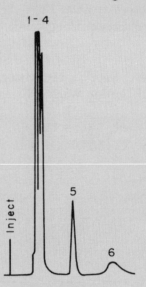

Sample : Six phthalate plasticizers

Fig. 8.3i. *Chromatogram requiring gradient*

Sample: Six phthalate plasticizers
1 R = CH_3 2 CH_2H_5 3 C_6H_5
4 n-C_4H_9 5 n-C_8H_{17} 6 iso-$C_{10}H_{21}$
Column: 10 μm C-18 bonded phase, 30 cm × 4 mm
Mobile phase: methanol/water 90 : 10
Flow rate 2 cm^3 min^{-1}
Detector: UV absorption, 254 nm

$\longrightarrow$

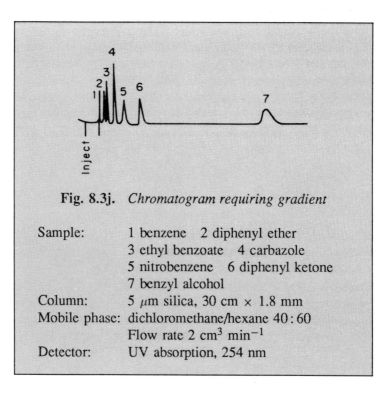

Fig. 8.3j. *Chromatogram requiring gradient*

Sample:	1 benzene 2 diphenyl ether
	3 ethyl benzoate 4 carbazole
	5 nitrobenzene 6 diphenyl ketone
	7 benzyl alcohol
Column:	5 μm silica, 30 cm × 1.8 mm
Mobile phase:	dichloromethane/hexane 40:60
	Flow rate 2 cm^3 min^{-1}
Detector:	UV absorption, 254 nm

Response

(*i*) For the reverse phase separation (Fig. 8.3i) the mobile phase needs to be more polar at the start and less polar at the end, so you should start with a mobile phase containing more water, say, methanol/water 75:25, and programme to pure methanol.

(*ii*) For the adsorption column (Fig. 8.3j) we want the mobile phase to be less polar at the start and more polar at the end, so we would start with a proportion of dichloromethane in hexane lower than 40% and finish with a concentration higher than 40%.

To sort out the best conditions, you would need to do some experimental work. Fig. 8.3k(*i*) shows the reverse phase separation run with the suggested gradient. In Fig. 8.3k(*ii*) the normal phase separation starts with 10% dichloromethane in hexane. This is run isocratically for 3 minutes, then the

proportion of dichloromethane is increased at 8% min⁻¹ to 40%. At this
point the programming rate was halved to 4% min⁻¹, and the programme
was continued to 100% dichloromethane.

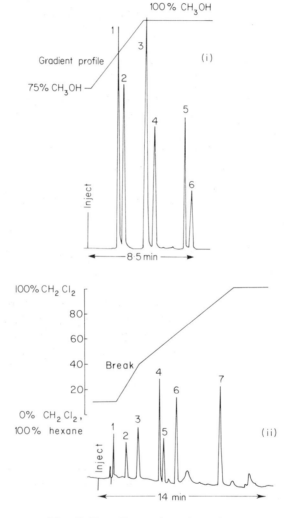

Fig. 8.3k. *Examples of gradients*

(*i*) Reverse phase separation of phthalates using a gradient
(*ii*) Normal phase separation using a gradient (sample as in Fig. 8.3j)

****

SAQ 8.4a	Fig. 8.4b is for the determination of benzoic acid and methyl and propyl parabens in sennoside. Sennoside is a laxative syrup and the three additives are used as stabilizers and preservatives. The parabens are alkyl esters of 4-hydroxybenzoic acid.

Chromatogram (i) is for a standard mixture of benzoic acid (0.06596 g dm^{-3}), methyl paraben (0.05802 g dm^{-3}) and propyl paraben (0.07470 g dm^{-3}) made up in the mobile phase.

Chromatogram (ii) is sennoside (256.4480 g dm^{-3}) made up in the mobile phase.

(a) What is the order of elution of the three solutes in Fig. 8.4b(i)?

(b) In Fig. 8.4b(ii), identify the peak for each additive in the mixture.

(c) From Fig. 8.4b(ii), using peak height measurement, calculate response factors for each additive and use these to calculate the concentration of each additive in the sample of sennoside.

Response

(a) In a reverse phase separation solutes are eluted in order of polarity, most polar first, so the elution order is benzoic acid, methyl paraben propyl paraben.

(b) For the standard chromatogram, retention distances are:

benzoic acid	43.5 mm
methyl paraben	76.5
propyl paraben	113

For the unknown, these three correspond to the peaks with retention distances of 43.5, 76.0 and 115 mm respectively.

(*c*)

	Peak height	
	standard	unknown
benzoic acid	36	14.5
methyl paraben	73	45.5
propyl paraben	88	10.5

	response factor	concentration of additive in sennoside
benzoic acid	$\dfrac{65.96}{36} = 1.832$	$14.5 \times 1.832 \times 3.899 = 103.6$ ppm
methyl paraben	$\dfrac{58.02}{73} = 0.795$	$45 \times 0.795 \times 3.899 = 141$
propyl paraben	$\dfrac{74.7}{88} = 0.849$	$10.5 \times 0.849 \times 3.899 = 34.8$

The factor $\dfrac{1000}{256.448} = 3.899$ allows for the dilution of the 'sennoside' before the analysis.

Injection no.		Aspirin	Phenacetin	Caffeine
	mass in mixture, g	0.6015	0.0765	0.0924
1	peak area	144 090	159 516	43 057
	relative response (*r*)	8.808	1	4.528
2	peak area	143 200	163 164	43 099
	r	9.066	1	4.627
3	peak area	121 297	139 796	36 564
	r	9.169	1	4.673
	Average *r*	9.014	1	4.609

Fig. 8.4f. *Completed*

****

SAQ 8.4b	A commercial analgesic tablet is stated on the packet to contain 325 mg of aspirin and 50 mg of caffeine per tablet. Two tablets plus 0.0773 g of phenacetin were shaken with 10 cm^3 of ethanol for 10 min, then 10 cm^3 of 0.5 mol dm^{-3} ammonium formate was added and the mixture was made up to 100 cm^3. The tablets contain insoluble excipients, so a little of the solution was filtered before chromatography. Fig. 8.4g gives peak areas obtained for two 1 μl injections of the solution. Chromatographic conditions were those described previously.

Injection no.	Peak area		
	Aspirin	Phenacetin	Caffeine
1	157595	170804	50693
2	153541	164174	48478

Fig. 8.4g.

Using Eq. 8.4b and the response factors from the last section, calculate the amount of aspirin and caffeine present, expressing the results as mg per tablet.

The aspirin content should be between 95 and 105% and the caffeine content between 90 and 110% of the amount stated on the packet. Are the tablets within specification?

Response

No. 1		Aspirin	Phenacetin	Caffeine
	peak area	157 595	170 804	50 693
	mass	0.6429		0.1057
	mg per tablet	321.4		52.9

For example, for aspirin $r = 9.014$, $A_u = 157\ 595$

$$\frac{c_s'}{A_s'} = \frac{0.0773}{170\ 804} = 4.526 \times 10^{-7}$$

$c_u = 157\ 595 \times 9.014 \times 4.526 \times 10^{-7} = 0.6429$ g present in two tablets $\equiv$ 321.4 mg per tablet

No. 2	Aspirin	Phenacetin	Caffeine
peak area	153 541	164 174	48 478
mass	0.6517		0.1052
mg per tablet	325.8		52.6

Alternatively we can correct each area and normalize the corrected areas, as follows.

No. 1	Aspirin	Phenacetin	Caffeine
peak area	157 595	170 804	50 693
r	9.014	1	4.609
corrected area	142 0561	170 804	233 644
normalized area %	77.83	9.36	12.80
mass	$\dfrac{0.0773}{9.36} \times 77.83 = 0.6428$	0.0773	0.1057
mg per tablet	321.4		52.9

The allowed limits are: aspirin 309–341 mg per tablet

caffeine 45–55 mg per tablet

Hence the tablets are within specification.

****

SAQ 8.4c

The following is an outline of a method for the determination of caffeine in decaffeinated instant coffee.

About 0.8 g of granules is accurately weighed, dissolved and made up to 50 cm^3. 5 cm^3 of this solution is shaken with 5 cm^3 of saturated lead acetate solution for 5 min. The resulting solution is then filtered before injection. A standard solution of caffeine, about 50 ppm, is also injected.

A typical chromatogram is shown in Fig. 8.1d.

(*a*) The sample solution injected should be dissolved in the mobile phase. How would you dissolve the granules to ensure this?

(*b*) Results from a typical determination are shown below:

mass of coffee taken	0.8277 g
caffeine standard: concentration	59.2 ppm
peak area	33 612
coffee solution: caffeine peak area	7262

From these results, calculate the percentage of caffeine in the coffee.

(*c*) A series of 10 determinations on the same sample of decaffeinated instant coffee gave a mean result of 0.149 ± 0.0046% for the caffeine level. Can you suggest any alterations to the experimental procedure that would improve the accuracy and the precision of the method?

Response

(*a*) The mobile phase is CH_3OH/H_2O 40:60 and also the coffee solution is shaken with an equal volume of aqueous lead acetate. The coffee could originally be dissolved in 80% CH_3OH.

(*b*) Response factor = $\dfrac{59.2}{33612}$ = 1.761×10^{-3}

Caffeine concentration in coffee solution

$$= 7262 \times 1.761 \times 10^{-3} = 12.79 \text{ ppm}$$

Original coffee solution contains $12.79 \times 2 = 25.58$ ppm (this allows for the dilution of the sample with lead acetate)

50 cm^3 contains $25.58 \times \dfrac{50}{1000} = 1.279$ mg caffeine

$\therefore$ % caffeine in the coffee = $\dfrac{1.279}{827.7} \times 100 = 0.155\%$.

(*c*) This method uses a caffeine external standard; the precision would be improved if an internal standard was used instead. Also, instant coffee is quite hygroscopic; drying the sample before analysis would certainly help. To look at the accuracy of the method we want to be sure that we have extracted all the caffeine—we could, for example, study the effect of dissolving the sample in hot water or simmering. We also need to know if the lead acetate extraction removes any caffeine, which we could investigate by using standards and subjecting them to the same procedure.

****

SAQ 8.4d	Draw up a table which lists briefly the advantages and limitations of each of the methods of quantitative analysis that were discussed in the previous section.

Response

A reasonable answer would look something like this:

Method	Advantages	Limitations
Area/height percent	No calibration necessary Injection volumes do not have to be precise Easiest and quickest method	All peaks in the sample must be eluted and resolved. Detector response must be the same for each component
External standard	One or more of the peaks can be determined Adjustment for components with unequal detector response	Calibration required Precision of injection volumes is essential
Internal standard	Includes all advantages of external standard, also compensates for small variations in injection volume and for small changes in detector sensitivity between runs	Calibration required Internal standard must be resolved from components in the sample; if added before sample preparation it should be chemically similar to components of interest also

****

SAQ 8.4e

Fig. 8.4h shows the UV spectra of four food additives. Tartrazine is a yellow colouring, saccharin and aspartame are sweeteners. Fig. 8.4i shows the chromatogram of a mixture of these additives, obtained using the following conditions:

Column: 5 μm C-18, 25 cm × 4.6 mm
Mobile phase: CH_3OH/NaH_2PO_4 pH 4.5 using a
               gradient from 5:95 to 95:5
Flow rate: 1.5 cm^3 min^{-1}
Detector: UV   (PDA)   using   wavelength
               programme

(*i*) Given that peak 3 on the chromatogram is caffeine, suggest the order of elution of the other solutes.

(*ii*) Suggest, with reasons, the probable outcome if the separation was done using isocratic elution with a mobile phase composition of, say, CH_3OH/NaH_2PO_4 20:80 or CH_3OH/NaH_2PO_4 80:20.

(*iii*) What problems would arise if a single wavelength of 254 or 280 nm was used for detection?

(*iv*) Suggest a suitable wavelength programme that could be used for the separation.

SAQ 8.4e
(cont.)

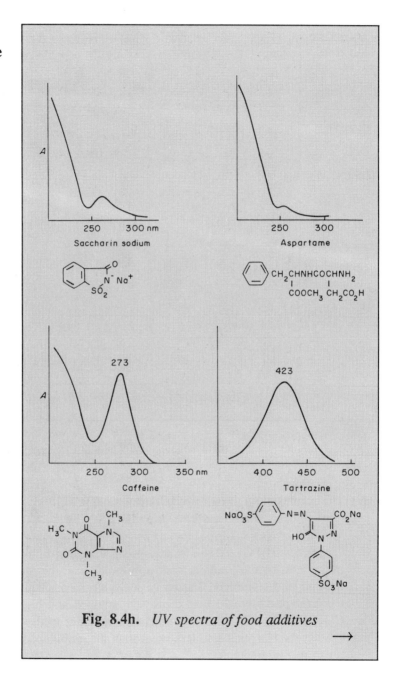

Fig. 8.4h. *UV spectra of food additives*

$\rightarrow$

SAQ 8.4e
(cont.)

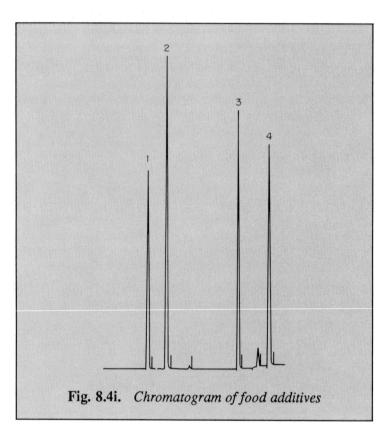

Fig. 8.4i. *Chromatogram of food additives*

Response

(*i*) The solutes elute in order of polarity, most polar first, so the order is tartrazine, saccharin, caffeine, aspartame.

(*ii*) CH_3OH/NaH_2PO_4 20 : 80 is less polar than the mobile phase used at the start of the gradient and more polar than the mobile phase at the end of the gradient. The early peaks would probably coelute, and the later peaks would be excessively retained and highly dispersed. CH_3OH/NaH_2PO_4 80 : 20 is much less polar than the mobile phase used at the start and also less polar than the mobile phase used at the finish of the gradient.

(*iii*) An operating wavelength of 254 nm would result in a low absorptivity for saccharin and a very low absorptivity for aspartame. 280 nm would result in a high sensitivity for caffeine, but would be unsuitable for saccharin and aspartame.

(*iv*) A suitable wavelength programme would be—peak 1 : 423 nm, peak 2: 210 nm, peak 3: 273 nm, peak 4: 210 nm.

****

9. Some Practical Aspects of HPLC

9.1. PACKING COLUMNS

Microparticulate stationary phases are packed into columns by forcing a slurry of the packing material in a suitable solvent into the column under high pressure. Manufacturers of columns are sometimes secretive about the methods they use for this, thus lending support to the view that the successful packing of HPLC columns is as much of an art as a science. If you need very high efficiency, or good reproducibility between columns, it is better to use manufactured columns. If not, you can make your own columns using simple and fairly inexpensive equipment. A number of suppliers sell systems for column packing, but it is much cheaper to buy the items individually and assemble them yourself.

Although there is no standard method for packing HPLC columns, there is a general consensus about the experimental conditions that are required. These are:

(*a*) For particle sizes below 10 μm spherical particles produce the best results; the smaller the particle size the more difficult it is to pack the column efficiently;

(*b*) The stationary phase particles must be properly dispersed in the slurry and must not coagulate;

(*c*) Sedimentation of the stationary phase should be avoided during the packing process;

(*d*) The stationary phase particles should hit the column bed with a high impact velocity;

(*e*) The bed should be packed under high compression; the pressure needed is greater the smaller the particle size of the packing;

(*f*) For small particle sizes, a low viscosity packing solvent is preferred.

Preliminary dispersion of the stationary phase in a suitable solvent is best carried out in an ultrasonic bath. To prevent sedimentation of the stationary phase during packing, a number of different approaches have been used:

(*a*) The stationary phase is dispersed in a solvent mixture which is formulated so as to have the same density as that of silica (2.2 g cm^{-3}). As one of the solvents must have a density greater than 2.2, the choice is rather restricted. For example, 1,1,2,2-tetrabromoethane (density 2.86) and tetrachloroethene (density 1.62) could be used.

∏   What composition would give the right density, assuming that the density of the mixture varies linearly with composition?

If v = volume fraction of $C_2H_2Br_4$, then

$$2.2 = 2.86v + 1.62(1 - v)$$

Hence, $C_2H_2Br_4 = 47\%$, and $C_2Cl_4 = 53\%$

Because the halogenated hydrocarbons that have to be used for this are both toxic and expensive, the use of balanced density slurries for packing columns is declining.

(*b*) Another method of reducing sedimentation in the packing slurry is to use a viscous solvent (e.g. glycerol/methanol mixtures).

∏   What do you think would be the difficulty with this method?

To achieve a reasonable flow rate through the column during the packing process [condition (*d*) above], very high pressures have to be used. Pressures greater than 1700 bar (25 000 psi) have been used for packing columns using viscous solvents. The apparatus has to be designed to withstand these high pressures, and consequently becomes expensive.

(c) A third approach is to use low density low viscosity solvents like methanol or propanone. With these, a satisfactory flow rate through the column during packing (about 15 cm³ min⁻¹ for 5 μm silica) can be obtained with relatively low pressures (roughly 350–650 bar, or 5000–10 000 psi). Sedimentation is minimized by not wasting time in those parts of the packing procedure where sedimentation can occur. In some methods, during packing, the slurry is contained in a reservoir fitted with a stirrer.

Fig. 9.1a shows the slurry packing system that I use. This operates at a fairly low pressure, as I am not especially interested in producing columns with very high efficiencies, but rather in saving on the cost of commercial columns. The pump and high pressure valve are rated for pressures of 500 bar (7500 psi) and 400 bar (6000 psi) respectively. The slurry reservoir is a stainless steel tube 85 cm long, with a capacity of about 50 cm³. The method is not hazardous unless there is air trapped in the high pressure line; nevertheless it is advisable to use a safety screen.

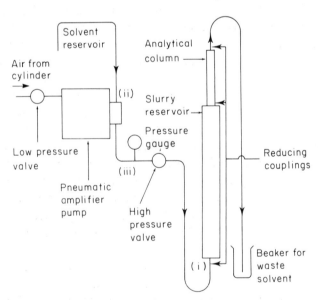

Fig. 9.1a. *Column packing system*

This is the method I use for packing a column with 5 μm bonded phase silica:

(*a*) The column tubing is washed with tetrachloromethane, then propanone, and finally dried. It is then fitted with a $\frac{1}{4}$ inch cap at the end that is to be attached to the slurry reservoir, and a 1/4–1/16 inch ZDV reducing coupling, fitted with a 1 μm stainless steel gauze, at the other end.

(*b*) For a 12.5 cm column, approximately 1.7 g of the silica is added to 30 cm^3 of methanol and the slurry is stirred with a magnetic follower until just before it is placed in the slurry reservoir. This quantity of silica is slightly more than is needed to fill the column.

(*c*) The packing solvent (400 cm^3 of methanol + 0.2 g of sodium ethanoate) is degassed under vacuum for 10 minutes and then placed in the solvent reservoir. The sodium ethanoate is added to prevent static build-up on the stationary phase during packing, which can produce unstable column beds, especially with bonded phase packings.

(*d*) With the low pressure valve closed, the air pressure is adjusted to about 6.7 bar (100 psi) using the second stage regulator on the air cylinder. The slurry reservoir is disconnected at point (*i*) and the pump is tested by opening the low pressure valve slightly and then opening the high pressure valve. If there is air trapped anywhere in the line the pump may not work at all, or the flow of solvent from point (*i*) may not be fast enough (this decision requires a bit of experience). If the solvent flow appears satisfactory, the high pressure valve is closed, whereupon the pressure on the Bourdon gauge should rapidly increase (to the value set by the inlet pressure and the amplification of the pump). If the pressure increases only slowly, this indicates the presence of air in the line. Air pockets are often present if the system has not been used for some time; to remove trapped air, the solvent line is disconnected at points (*ii*) and (*iii*) and solvent is passed rapidly through each stage, using a 20 cm^3 syringe with a 1/16 fitting attached to the needle.

(*e*) Assuming the pump is behaving properly, the solvent line is connected at point (*i*), the high pressure valve is closed and the system is pressurized to 350 bar (about 5000 psi) by opening the low pressure valve. The remaining operations in (*e*) are carried out as quickly as possible, to limit the effects of sedimentation. The slurry is poured into the reservoir, which is then topped up with methanol. The column and outlet tube are connected and then the high pressure valve is opened.

(*f*) After about 200 cm^3 of solvent have been pumped through the column (about 15 minutes), the column and slurry reservoir are inverted, pumping is continued for a further 5 minutes, then the high pressure valve is closed and the pressure on the pump side is released by closing the low pressure valve. After 10 minutes the column is disconnected from the reservoir, the top of the packing is smoothed with a razor blade and a gauze and reducing coupling are fitted.

(*g*) The column is then attached to the injection unit on the chromatograph (but not to the detector) and mobile phase is pumped through the column for 10 minutes at about 3 cm^3 min^{-1}. The column is then connected to the detector and pumping is continued at a lower flow rate until (with a UV absorbance detector) a stable baseline can be obtained on the lowest absorbance setting.

When the column is ready to be used, the chromatogram of a suitable test mixture should be obtained. The plate number and retention times of the test solutes should be noted, and the peaks should have a satisfactory shape (minimal tailing). For measurement of the plate number, the recorder should be used at a high chart speed. Fig. 9.1b shows a test chromatogram for a C-18 column prepared by the above method, and Fig. 9.1c and 9.1d show the data that you should report with the chromatogram. The retention for an unretained peak is taken as the small baseline disturbance just before the first peak.

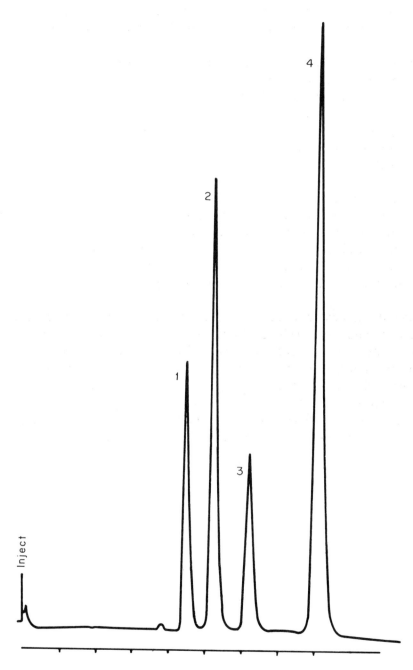

Fig. 9.1b. *Chromatogram of a test mixture*

Column length	12.5 cm
Internal diameter	4.6 mm
Stationary phase	5 μm C-18
Mobile phase	CH_3OH/H_2O 60:40
Flow rate (nominal)	1 cm^3 min^{-1}
Pressure drop	83 bar (1250 psi)
Temperature	ambient (about 22 °C)
Detector	UV absorption, 254 nm, 0.5 a.u.f.s.
Injection volume	0.5 μl
Recorder	10 mV, 10 mm min^{-1}
Test mixture peaks (1–4)	propanone, phenol, 4-hydroxymethylbenzene, methyl phenyl ether, dissolved in mobile phase.
Unretained solute retention distance	38 mm

Fig. 9.1c. *Conditions used for chromatogram in Fig. 9.1b*

Π   (*a*) See if you can complete the data in Fig. 9.1d from measurements on the chromatogram. The correct values are given at the end of the section.

Peak	1	2	3	4
Retention distance, mm		52		81
k'		0.37		1.13

Fig. 9.1d. *To be completed.*

(*b*) Measure the width of peak 4 at the baseline and calculate the plate number and plate height of the column (Eqs. 2.2a–c).

$$w = 4.5 \text{ mm}, N = 16 \times \frac{81^2}{4.5^2} = 5184, H = 0.024 \text{ mm}$$

It would be better to do this measurement from a chromatogram recorded at a higher chart speed, so that the peak width could be measured more accurately.

The column was satisfactory apart from the pressure drop, which was high, indicating a partial blockage, probably of one of the gauzes. In order to measure k' for the solutes we have to know the retention distance or volume of an unretained solute. The accurate determination of this quantity is not an easy problem with bonded phase columns and reverse phase operation. A number of different methods have been used:

(*a*) Determination of the difference in mass between the column full of solvent and the dry column. As dry columns do not easily reconstitute, this procedure is best done towards the end of the life of the column;

(*b*) Injection of a homologue of one of the mobile phase components, with detection by refractive index. For example, with a methanol/water mobile phase, ethanol could be used;

(*c*) Injection of D_2O or labelled components of the mobile phase;

(*d*) Mathematical methods based on retention data obtained for a homologous series of compounds;

(*e*) Injection of unretained solutes, e.g. $NaNO_3$ or tartrazine, which can be detected by UV absorption. The idea is that such highly polar solutes will not be retained by the non-polar stationary phase. Sometimes, the flow disturbance before the first peak is used (as above), or even the first peak itself is taken as an unretained solute if it is a very polar compound.

The trouble is that the various methods all give different results. In particular, the gravimetric method usually gives higher results than the others. The gravimetric method will measure the total volume of solvent in the column, i.e. void volume + interstitial volume. Lower results will be obtained from the injection of solutes if the solutes are partly excluded from the pores of the stationary phase. If the measured retention distance from the chromatogram is converted into a volume, the flow rate from the pump must be carefully determined by collecting the mobile phase for a known time and weighing it.

Peak	1	2	3	4
Retention distance, mm	45	52	62	81
k'	0.18	0.37	0.63	1.13

Fig. 9.1d. *Completed*

9.1.1. Summary

Techniques for packing HPLC columns and for testing the packed column are described.

Learning Objectives

You should now be able to:

• Appreciate the experimental conditions needed for the successful packing of HPLC columns;

• Describe how a column is packed;

• Understand how to evaluate a test chromatogram.

9.2. THE PREPARATION OF MOBILE PHASES

This section describes some of the problems that can occur with the mobile phase in HPLC. Many of these problems arise because of the presence of impurities, additives, dust or other particulate material, or dissolved air. It is always best to try to prevent these problems by a little attention to detail and the use of simple good housekeeping procedures. Although it is always tempting to try to save time and expense by the neglect of such matters, if you do this you will store up trouble for yourself in the long term.

A major cause of practical problems in HPLC is the presence of air bubbles in the mobile phase at some point in the system. Some of the symptoms of trapped air were dealt with in Section 3. Air bubbles can collect in the pump, or the detector cell, or in other places. Because of their compressibility, air bubbles will reduce the volume of mobile phase delivered by the pump so that reproducibility is affected, and because of the flow variation the detector noise is often worse as well. Large air bubbles in the pump may stop the pump working. Detection can be affected in various ways. For example, with UV absorbance detectors, air in the detector cell can cause serious noise, or very high absorbance. Dissolved oxygen can interfere with UV absorbance detection at short wavelengths, as oxygen absorbs radiation strongly below 200 nm. Many problems with dissolved air are avoided if the mobile phase is degassed before use, so this should always be done. Degassing can be accomplished by placing the mobile phase under vacuum, or by heating and ultrasonic agitation, stirring, or by a combination of these. In many practical arrangements, having removed the air, the mobile phase is then placed in a reservoir in contact with air, allowing the uptake of air to start again. If this is the case, degassing should be repeated every hour or so. To restrict access of air to the mobile phase, some arrangements use a straight sided reservoir with a plastic float that sits on top of the mobile phase. Another technique involves saturating the mobile phase with helium, which has a small solubility in liquids. Access of air is restricted by having a continuous slow stream of helium passing over the mobile phase in the reservoir. Although helium is rather expensive, only a small amount is used, and this method is now very popular.

A microparticulate HPLC column is a very efficient filter, and if the mobile phase contains any particulate matter, or acquires it from the pump and/or the injection valve due to wear, it will collect at the top of the column. If this happens, the pressure drop across the column for a given flow will gradually increase, and the column may eventually become completely blocked. To prevent this happening, the mobile phase should always be filtered before use, preferably through a 0.5 μm porosity filter, and guard and scavenger columns should be used as a matter of routine (see Section 9.3.2).

Many reagent grade solvents contain levels of impurities that make them unsuitable for long term use in HPLC. Sometimes the impurities are added deliberately as antioxidants, stabilizers, or for denaturing. Wherever possible, 'HPLC grade' solvents should be used to prepare mobile phases, or alternatively the solvents should be adequately purified before use.

Distilled or deionized water contains small amounts of organic impurities, which can cause problems in long term use with bonded phase columns in the reverse phase mode. The non-polar stationary phase will collect these organics, which can alter the nature of the stationary phase or sometimes produce spurious peaks (Fig. 8.3c is an example of this). High purity water degrades in quality during storage, even in glass containers, and for trace level work the water should be purified at the time of use. Water purification can be done by distillation from permanganate, by passage of the water through bonded phase columns or by means of commercial systems; e.g. the Millipore system uses a combination of activated carbon adsorption, mixed bed deionization, an organics-scavenging cartridge and a 0.22 μm sterilizing filter at the outlet. This produces water free of contaminating ions and containing less than 15 ppb of total organic carbon.

Impurities in other solvents may affect chromatographic behaviour, or detection, or both. Chlorinated solvents such as di- or trichloromethane are stabilized against oxidative breakdown by the addition of small amounts of methanol or ethanol.

∏   In a normal phase separation using a trichloromethane/heptane mobile phase, how would the presence of stabilizer in the trichloromethane affect the separation?

The presence of alcohol would increase the polarity of the mobile phase, so that solute retention times would be shortened. Also, we would not expect to get very good reproducibility, as the concentration of stabilizer will vary slightly from batch to batch.

Chlorinated hydrocarbons can be purchased without stabilizer, or the stabilizer can be removed by adsorption on to alumina, or by extraction with water, followed by drying. Unstabilized di- or trichloromethane will slowly decompose, producing HCl, which corrodes stainless steel. The rate of decomposition may be accelerated by the presence of other solvents.

Ethers contain additives to stabilize them against peroxide formation. For instance, tetrahydrofuran is commonly stabilized by the addition of small amounts of hydroquinone. This absorbs UV radiation strongly and so interferes with UV absorbance detection. It can be removed by distilling the solvent from KOH pellets. If you use inhibitor-free tetrahydrofuran, it should be stored in a dark bottle and flushed with nitrogen after each use. Any peroxides that form should be periodically removed by adsorption on to alumina.

When mixing solvents to form mobile phases, the volume of each component should be measured separately before the solvents are mixed, since the volume of the mixture does not usually equal the sum of the two separate volumes. For example, 50 cm^3 of methanol mixed with 50 cm^3 of water produces a total volume of about 96 cm^3 of a 1 : 1 mixture. If you make up a mobile phase by filling a measuring cylinder half full of methanol and then making up to the mark with water, you won't end up with a 1 : 1 mixture. If the mobile phase contains volatile components, the composition can alter during degassing procedures. Measurement of the refractive index of the mobile phase is a useful check on the composition.

With UV absorbance detectors, we have to consider the UV absorption of the mobile phase, which always increases as the wavelength decreases. The 'UV cut-off' of solvents indicates the useful wavelength range of the solvent and means the wavelength below which the solvent has an absorbance of 1 or more when measured in a 1 cm cell. Aliphatic hydrocarbons cut off at about 210 nm, and the best polar solvents for low wavelength work are water, methanol and acetonitrile. The last two cut off at 205 and 190 nm respectively, provided they are pure. Acetonitrile is difficult to purify and is consequently expensive.

Other properties of HPLC solvents that we may need to consider include compressibility, viscosity, refractive index, vapour pressure, flash point, odour and toxicity. Most HPLC textbooks contain tables of these properties. For instance, there is a useful table in the book edited by J.H. Knox.

The solvent properties listed in Fig. 9.2a are used in SAQ 9.2b.

	Viscosity cp, 20 °C	Boiling point, °C	UV cut-off, nm	Price	TWA*, ppm
pentane	0.23	36.2	210	12.76 (*i*)	500
heptane	0.43	98.4	210	13.20	500
trichloromethane	0.57	61.2	245	7.80	50
tetrachloromethane	0.97	76.8	265	9.20 (*i*)	10
acetonitrile	0.37	82.0	190	10.30	40
dioxane	1.54	101.3	220	14.40	100
methanol	0.60	64.7	205	4.60	200
ethanol	1.20	78.5	210	21.40 (*ii*)	1000
propane-2-ol	2.30	82.3	210	7.68	200
propanone	0.56	56.5	330	5.20	1000

Fig. 9.2a. *Properties of solvents commonly used in HPLC*

Prices are given in £/dm^3 (1990) for BDH HiPerSolv HPLC solvents, except for:

(*i*) AnalaR reagent; (*ii*) 99.7–100% AnalaR reagent, less duty.

* Time weighted average (TWA) is the average contamination level to which a worker may be exposed continuously for eight hours without damage to health. The values were taken from *SKC Guide to Air Sampling Standards*, 1989.

SAQ 9.2a

The following account, taken from a practical notebook, describes the preparation of the mobile phase used in the example in Section 8.4.3, to which you will need to refer.

'1.575 g of ammonium formate was made up to 500 cm^3 in a graduated flask. To this solution, 50 cm^3 of ethanol was added, and after mixing the mobile phase was placed in the solvent reservoir and pumping was commenced at 2 cm^3 min^{-1}'.

Can you identify three mistakes that were made?

SAQ 9.2a

SAQ 9.2b | These questions refer to the properties of solvents that were listed in Fig. 9.2a.

(*i*) Considering a heptane/trichloromethane mobile phase, why would it be undesirable to replace the heptane by pentane, or the trichloromethane by tetrachloromethane?

(*ii*) Why is methanol or acetonitrile preferred to other water miscible solvents for the preparation of mobile phases for reverse phase chromatography?

(*iii*) What would be the major difficulty associated with the use of propanone in mobile phases?

SAQ 9.2b

SAQ 9.2c You are performing a separation using the following conditions:

Column: 5 μm C-18, 25 cm × 4.6 mm
Mobile phase: acetonitrile/phosphate buffer, pH
                   6.9, 75 : 25
Detector: UV absorption, 230 nm
Flow rate: 2 cm^3 min^{-1}
Temperature: ambient
Injection volume: 25 μl, using autosampler

$\longrightarrow$

				Retention time (min)		
Day	Time	t_0	Peak 1	Peak 2	Peak 3	
1	1000	1.25	2.17	12.12	17.84	
	1300	1.26	2.19	12.17	17.82	
	1600	1.24	2.18	12.15	17.85	
2	0900	1.24	2.36	13.03	19.19	
	1100	1.26	2.35	13.01	19.22	
	1300	1.25	2.38	13.06	19.21	

SAQ 9.2c
(cont.)

The retention times that you get for the peaks vary from day to day; the table below shows some typical results.

Make a list of the more important experimental variables that are likely to affect retention in a reverse phase separation. Using the retention data above, is it possible to suggest which of these variables might be the cause of the problem?

9.2.1. Summary

Problems can occur with the mobile phase because of the presence of particulate matter, impurities or dissolved air. Some of the practical remedies are considered.

Learning Objectives

You should now be able to:

• Appreciate the need to use mobile phases of sufficient purity;

• Recognize some of the symptoms of the presence of air or other impurities in the mobile phase;

• Identify the methods used to degas mobile phases.

9.3. PRACTICAL PROCEDURES WITH COLUMNS AND SAMPLES

9.3.1. Care of HPLC Columns

The useful lifetime of HPLC columns is shortened by the appearance in the column packing of cracks or voids, especially at the top of the column, or by the collection at the top of the column of particulate material from the mobile phase (e.g. wear particles from the pump or injection valve), or strongly held components from the sample.

The appearance of voids at the top of the column bed is a common problem, which can be caused by the gradual settling of the column packing or by the dissolution of the silica stationary phase, caused, for instance, by the use of alkaline mobile phases. The presence of a void at the top of the column will result in a loss of efficiency because of the extra dead volume in the system. If the profile of the column bed at the top of the column is irregular, a single solute may produce a peak with a shoulder, a double peak or even a group of peaks. What happens is that, as the solute enters the column, it encounters the stationary phase at different points in time, and the end result is as though two or more chromatograms are superimposed, all slightly out of phase. The effect is shown in Fig. 9.3a; Fig. 9.3b shows a practical example of this type of behaviour.

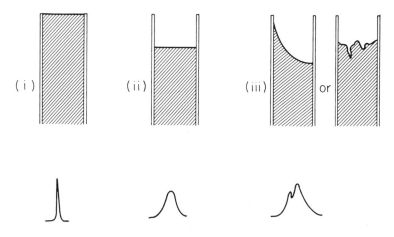

Fig. 9.3a. *Packing faults at the top of the column*

A void at the top of the column, (ii), produces additional dispersion. Uneven packing as in (iii) adds to the dispersion and may cause a solute to appear as two or more peaks

The packing at the top of the column can be repaired by first removing the inlet compression fitting and frit. A small quantity of the packing is then scraped out with a small screwdriver or micro-spatula, then the packing is tamped down gently with a glass or plastic rod of the right diameter. The void can be filled with ballotini (glass microspheres) or a thick slurry of the stationary phase in a suitable solvent. The packing is levelled at the top with a razor blade, and then the frit and compression fitting are replaced.

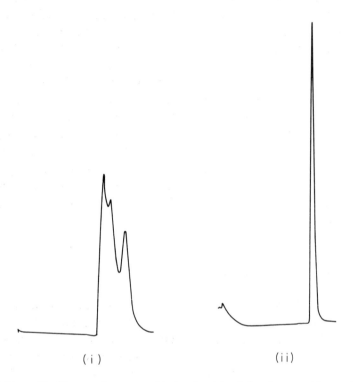

(i)   (ii)

Fig. 9.3b.   *(i) Chromatogram obtained with defective packing at the top*
of the column

Column: 5 μm C-18 12.5 cm $\times$ 4.9 mm
Mobile phase: CH_3OH/H_2O 60:40
Flow rate: 1 cm^3 min^{-1}
Detector: UV absorption, 254 nm
Sample: propanone

(ii) The same sample after repair of the top of the column

Although HPLC columns are packed under high pressure, the column bed
can be disturbed by sudden pressure variations, or mechanical or thermal
shock, so to extend the lifetime of columns all of these should be avoided.
Columns should be taken gradually from low to high or from high to low
pressures and sudden temperature changes should be avoided. They should
be stored where they are unlikely to be bumped or jarred. When not in

use, columns should be stored in an inert non-volatile solvent. Usually the storage solvent should have the same polarity as that normally used with the column. For instance, bonded phase columns for use in reverse phase mode can be stored in methanol or methanol/water mixtures. These are usually preferred to distilled water, as the methanol prevents the growth of bacteria. The columns should be securely capped at each end, preferably with compression fittings. Always make a note of the storage solvent that is used; if you do not, you may have miscibility problems the next time you use the column. For instance, if a column is stored in methanol and is subsequently used with an aqueous buffer, it is necessary first to wash the column with water, otherwise the buffer salts may precipitate when they encounter the methanol. Stainless steel columns and fittings are slowly corroded by buffer solutions, which should always be removed from columns after use.

9.3.2. Protection of Columns During Use

The useful life of an analytical column is substantially improved by the use of pre-columns, located between the pump and injector (this sort are sometimes called *scavenger* columns) and/or between the injector and the analytical column (*guard* columns). The scavenger column is an efficient final filter that protects the analytical column from wear particles from the pump and from dust or other particulate matter in the mobile phase. Guard columns protect the analytical column from wear particles from the injection valve and from any adverse characteristics of the sample. Complex samples may contain components that are irreversibly held on the stationary phase under the conditions of the separation. If these materials build up at the top of the column they can have a profound effect on chromatographic behaviour such as retention, selectivity and efficiency. What we are doing with guard and scavenger columns is to transfer these problems from a relatively expensive analytical column to a relatively cheap pre-column.

Because scavenger columns are located upstream of the injection valve, they do not add to the dispersion of the chromatogram, and their size is not critical in this respect. Guard columns, on the other hand, do cause a slight loss of efficiency, and so need to have a relatively small volume. Reducing the volume, of course, reduces the life of the guard column.

For use with 25 cm × 4.6 mm analytical columns, guard columns and scavenger columns are often 4.6 mm in internal diameter and 3–10 cm in length. They can be packed with microparticulate stationary phases or with porous layer beads. Porous layer beads are cheaper than microparticulates and are easier to pack, but they have lower capacities and will require changing more often. It is usually difficult to know how long a pre-column will last before it requires changing. In routine work, pre-columns are usually repacked or replaced to a fixed schedule.

An alternative use of pre-columns is in sample pre-concentration steps. Solutes present at very low levels in a sample can sometimes be concentrated by passing a large volume of sample through a small column on which the solutes are strongly retained. This column can then be connected to the analytical column and the solutes eluted with a suitable mobile phase. The pre-column can be packed with a stationary phase having a relatively large particle size so that the sample can be pumped through rapidly at low pressure, using a cheap pump. Pre-columns filled with high-porosity silica have also been used in work with basic mobile phases to extend the lifetime of the analytical column by saturing the mobile phase with silica.

If the performance of a column is no longer satisfactory, it can sometimes be reconditioned by washing with a suitable solvent, or series of solvents. Some bonded phase columns, C-18 for instance, tend to collect non-polar impurities, which can sometimes be removed by washing the column with a non-polar solvent, e.g. heptane. Assuming the mobile phase normally used with the column is CH_3OH/H_2O 50:50, we cannot wash directly with heptane; because of miscibility problems, we have to get to heptane via a miscible solvent or series of solvents.

Π   How could this be done?

> You could first wash the column with methanol, then trichloromethane, then heptane (or methanol, ethyl ethanoate, heptane). You cannot go directly from methanol to heptane because the two are only partly miscible. The column needs to be washed with about 20 dead volumes of each solvent (about 50 cm^3 of each solvent for a 25 cm × 4.6 mm column). To get back to CH_3OH/H_2O 50:50 you would have to go through the sequence of solvents in reverse. If buffer solutions or ion-pairing reagents have been used

in the mobile phase, very much longer equilibration times may be needed.

9.3.3. Sample Preparation and Clean-up

Samples in HPLC can come from a very wide range of sources, and unfortunately they are usually not simple mixtures of pure organic compounds, all of which are soluble in and separable by the same mobile phase. Biological samples may contain proteins, salts and a host of organic compounds with widely differing polarities. Pharmaceutical samples often contain a wide range of soluble and insoluble excipients. Samples to be analysed for environmental pollution can and do contain almost anything! Very often, we are concerned with the separation of sample components of interest, sometimes present at very low levels, from a range of other components in the sample matrix which may interfere with the analysis.

Many of the adverse consequences of injecting dirty samples can be prevented or minimized by the use of guard columns, as discussed earlier, but often some form of sample clean-up is needed as well. The goal of sample preparation is to obtain, from the sample, the components of interest in solution in a suitable solvent, free from interfering constituents of the matrix, at a suitable concentration for detection and measurement. Naturally we want to do this with the minimum time and expense.

Whenever possible, the sample should be dissolved in the same solvent or mixture that is used for the mobile phase. If a different solvent is used, this often results in loss of efficiency and poor peak shape. What happens is that, as the injected material moves down the column, sample molecules at the edges of the injected band are in contact with a mobile phase that has a different composition to that seen by the bulk of the sample. The result is that the molecules at the band edges will travel at a different speed to that of the rest of the sample, resulting in spreading or splitting of the peaks.

Fig. 9.3c shows the consequences of dissolving a sample in solvents that are not the same as the mobile phase.

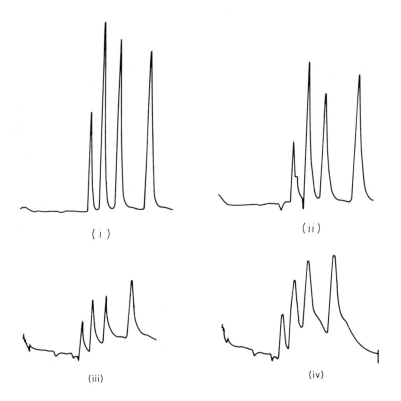

Fig. 9.3c. *Injection of a test mixture dissolved in different solvents*

Column: 5 μm C-18 bonded phase 12.5 cm × 4.9 mm
Mobile phase: CH_3OH/H_7_2O 60 : 40
Flow rate: 1 cm^3 min^{-1}
Detector: UV absorption, 254 nm
Sample: as in Fig. 9.1b, dissolved in (*i*) mobile phase, (*ii*) tetrahydrofuran, (*iii*) ethanol, (*iv*) butanol

Traditional methods of sample preparation such as liquid–liquid or liquid–solid extraction are time consuming and can sometimes result in incomplete sample recovery. They have to a large extent been replaced by column extraction procedures. In many ways, these resemble the use of guard columns, except that the extraction or purification is done before the chromatography. There are a number of commercial systems for column extraction, e.g. the Waters 'Sep-Pak' cartridges, which are radially compressed plastic 'mini-columns', 2.5 cm long × 1 cm diameter, through which samples can be passed using a syringe.

These are available packed with silica, alumina, C-18 bonded silica, or silica bonded with a variety of other chemistries. With these, you can either choose conditions (cartridge and solvent) such that the sample components of interest are retained on the packing while the matrix interferences pass through unretained, or you can retain the interfering components. The second approach is usually chosen when the components of interest are present at relatively high levels. When the components of interest are retained on the cartridge they can subsequently be removed by eluting with a solvent of a different polarity. Fig. 9.3d shows the principle of the method, and two examples of its use are given in the SAQs at the end of this Section.

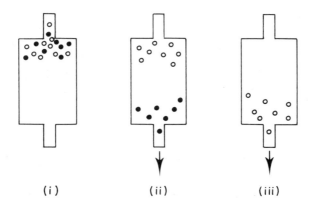

Fig. 9.3d. *Operation of a C-18 extraction cartridge*

(*i*) Load sample
(*ii*) Elute with polar solvent (e.g. H_2O) to remove polar material
(*iii*) Elute with less polar solvent (e.g. CH_3OH) to remove non-polar material.

An interesting short cut to the problem of sample preparation is provided by the Pinkerton ISRP (internal surface reverse phase) column concept. The packing in these columns is based on a 5 μm spherical silica modified to a hydrophilic external surface but with the internal surface (the pores) having a selective reverse phase bonding. They are used for the determination of drugs or other analytes in serum. Proteins in the serum are not retained by the outer surface of the packing and are too large to get into the pores of the structure. They are therefore unretained and pass straight through the column. Drugs and other metabolites, being smaller, can penetrate the inner surface of the packing and undergo a reverse phase separation. Conventional sample preparation in such cases might have involved precipitation of the protein, solvent extraction of the drugs from the supernatant liquid, evaporation of the extracting solvent, dissolution of the residue in mobile phase and filtration.

The following SAQs involve two examples of the use of solid phase extraction methods. First we will consider a recent publication on the determination of some two and three ring azines in water (T.R. Steinheimer and R.G. Ondrus, *Analytical Chemistry* 1986, 58, 1839–1844).

The compounds of interest are nitrogen-containing analogues of polycyclic aromatic hydrocarbons (PAHs) and include such substances as quinoline (1-benzazine) and acridine (2,3,5,6-dibenzopyridine). They are thought to be released into the environment as a result of combustion or pyrolysis processes involving fossil fuels. Some of them are highly carcinogenic. In contaminated water samples they are usually accompanied by PAHs, which are present at much higher levels. The azines in contaminated water samples were first enriched and separated from accompanying PAHs using an extraction technique on a non-polar cartridge. Details of this, and of the conditions used in the chromatography, are summarized below. After reading the summary, see if you can answer the problems in SAQ 9.3a.

A sample of water with a volume of up to 2 dm^3 was filtered through Whatman No. 1 filter paper to prevent clogging of the extraction cartridge. The filtered water sample was then passed through a Waters C-18 Sep-Pak at flow rates of up to 200 cm^3 min^{-1}. The cartridge was centrifuged to remove the residual water and was then eluted with 2 cm^3 of acetonitrile/0.78 mol dm^{-3} HCl 25:75. The eluate was neutralized with aqueous ammonia and then filtered through a 0.2 μm porosity nylon filter.

For the chromatographic separation of the azines, the column used was 10 cm × 8 mm Nova-Pak (4 μm C-18 bonded phase). The mobile phase was acetonitrile/water 42:58, pH 7.2, pumped at 1.5 cm^3 min^{-1}.

SAQ 9.3a

These questions refer to the extraction and separation of azines that was discussed in the previous Section:

(*i*) What difference between the azines and the PAHs is exploited to achieve their separation on the C-18 cartridge?

(*ii*) Why is HCl used when the cartridge is eluted?

(*iii*) If the PAHs had not been removed from the sample, how would you expect their retention times to compare with those of the corresponding azines (e.g. anthracene and acridine or naphthalene and quinoline)?

(*iv*) The chromatographic separation was done at pH 7.2. Can you think of any disadvantages of using a lower or a higher pH?

(*v*) Which detector could be used for the separation?

The next problem refers to the determination of oxalic acid (ethanedioic acid) and malic acid (hydroxybutanedioic acid) in rhubarb. (B. Libert, *Journal of Chromatography* 1981, 210, 540–543). Oxalic acid is present in rhubarb both as the free acid and as calcium oxalate, which is insoluble in water. The pK_a values of the two acids are shown below:

   oxalic acid: $pK_{a1} = 1.23$, $pK_{a2} = 4.19$
   malic acid: $pK_{a1} = 3.40$

A short summary of the method used is described below. About 25 g of frozen rhubarb stalks were homogenized and extracted in 200 cm^3 of 1 mol dm^{-3} HCl for 15 min at 100 °C on a water bath. The solution was stored overnight and, after filtration and dilution, 4 cm^3 of the diluted filtrate was passed through a C-18 extraction cartridge. The first 2 cm^3 were discarded and the rest was used for analysis. Chromatographic conditions were:

   Column: LiChrosorb 10 μm C-8, 25 cm × 4.6 mm
   Mobile phase: KH$_2$PO$_4$ 0.0367 mol dm^{-3}, tetrabutylammonium hydrogen sulphate (TBA) 0.005 mol dm^{-3} buffered to pH 2 with H$_3$PO$_4$
   Detector: UV absorption, 220 nm

SAQ 9.3b

These questions refer to the determination of dibasic acids in rhubarb, discussed in the previous section.

(*i*) How is the precipitation of calcium oxalate prevented?

(*ii*) What would be the order of elution of the two acids if the TBA was not present in the mobile phase?

(*iii*) The TBA is an ion pairing reagent. What effect will it have on the retention of the two acids?

(*iv*) What is the purpose of the C-18 extraction?

SAQ 9.3b

SAQ 9.3c

The following are all common practical problems when using HPLC. Suggest a probable cause of the problem and a remedy in each case.

(*i*) The pump motor is operating, but the pressure drop across the column suddenly falls to zero, and the flow of mobile phase stops.

(*ii*) Using a UV detector, there is a sudden upward step in the absorbance, or a series of large noise spikes on the chromatogram, or both effects at once.

(*iii*) Using a UV detector at 195 nm, the baseline is noisy, absorbance is high, and peak heights show poor reproducibility.

(*iv*) The pressure drop across the column increases until the cutout limit is reached, when the pump switches itself off.

SAQ 9.3c

9.3.4. Column Switching

The use of extraction cartridges is one example of a technique known as column switching (it is also often called *multidimensional* chromatography). The method can be used, either on-line or off-line, for sample clean-up by selecting part of a complex chromatogram (a *cut*) and transferring the cut to one or more secondary columns for further separation. Alternatively, column switching can be used for on-column concentration, in which a large volume of sample is passed through a pre-column under conditions such that the solutes of interest are retained. After concentration in this way, the mobile phase is changed so that the solutes are eluted rapidly, and another column is used for the analysis. The use of extraction cartridges in the separation of azines, discussed in the last Section, is an example of on-column concentration using off-line column switching. A chromatogram can be cut off-line by collecting the zones of interest at the detector outlet followed by reinjection of the collected fraction on to a second column. The mobile phases used with the two columns should be compatible, e.g. they should be miscible and the mobile phase used with the first column should not have too high an eluting power in the second column. If the mobile phases are incompatible it may be possible to evaporate the primary mobile phase and redissolve the sample in a suitable solvent.

The example shown in Fig. 7.7g uses off-line column switching to combine exclusion and reverse phase chromatography for the separation of pesticides from a complex sample matrix.

With on-line techniques, the column switching operations are done using valves. Fig. 9.3e shows a simple arrangement for zone cutting that could be used for sample clean-up. The zone marked Y is to be determined and all other zones are to go to waste (this type of cut is called a *heart cut*). Initial separation takes place on column C1 so that early zones (X) are routed to waste. When zone Y is eluted from C1, valve V2 is switched to elute this zone on to column C2. After complete transfer of Y on to C2, valve V1 is switched to prevent further elution of unwanted zones (Z, for instance). Zone Y is eluted to the detector and C1 can be cleaned and re-equilibrated with mobile phase.

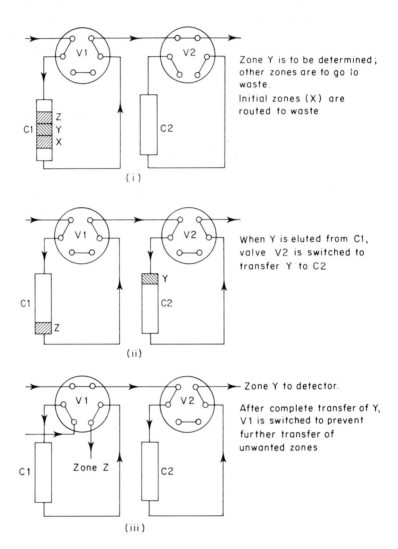

Fig. 9.3e. *Column switching*

SAQ 9.3d   Comparing on-line and off-line methods of column switching, which of the following advantages and limitations do you think would apply to the on-line technique?

   (*i*) Easier to carry out

   (*ii*) Cheaper

   (*iii*) Faster

   (*iv*) Better reproducibility

   (*v*) More chance of sample loss

9.3.5. Summary

The useful life of HPLC columns can be extended by proper treatment, in particular by the use of guard and scavenger columns. Pre-treatment of samples is often necessary, e.g. very dilute samples may require concentration, or complex samples may need to be cleaned up. Some of the more important techniques are considered.

Learning Objectives

You should now be able to:

- Appreciate some of the conditions needed to prolong the life of HPLC columns;

- Understand the function of guard columns and scavenger columns in protecting the analytical column;

- Recognize that column switching techniques can be useful in the treatment of dilute or complex samples.

References

PACKING COLUMNS

1. J.H. Knox, Ed. *High Performance Liquid Chromatography*, Edinburgh University Press, 1982, Chapter 12.

2. M. Verzele and C. Dewale, *LC–GC* 1986, 4(7), 614–618.

OTHER PRACTICAL METHODS

3. D.J. Runser, *Maintaining and Troubleshooting HPLC Systems*, Wiley–Interscience, 1981.

4. C. Gertz, *HPLC Tips and Tricks*, Lab. Data Control Ltd.

5. F.M. Rabel, *Journal of Chromatographic Science* 1980, 18 394–408.

6. C.F. Simpson, Ed. *Techniques in Liquid Chromatography*, Wiley, 1984, Chapter 3.

7. R.E. Majors, *LC–GC International*, 1991, 4(2), 10–14.

8. J.W. Dolan and L.R. Snyder, *Troubleshooting LC systems*, Humana, Clifton, NJ, 1989.

9. C. Markell, D.F. Hagen and V.A. Bunnelle, *LC–GC International* 1991, 4(6), 10–14.

COLUMN SWITCHING

10. C.J. Little, O. Stahel, W. Linder and R.W. Frei, *American Laboratory* 1984, 16(10), 120–122, 125–129.

11. R.E. Majors, *Journal of Chromatographic Science* 1980, 18, 571–579.

12. M.J. Koenigbauer and R.E. Majors, *LC–GC International* 1991, 3(9), 10–16.

SAQs AND RESPONSES

SAQ 9.2a

The following account, taken from a practical notebook, describes the preparation of the mobile phase used in the example in Section 8.4.3, to which you will need to refer.

'1.575 g of ammonium formate was made up to 500 cm^3 in a graduated flask. To this solution, 50 cm^3 of ethanol was added, and after mixing the mobile phase was placed in the solvent reservoir and pumping was commenced at 2 cm^3 min^{-1}'.

Can you identify three mistakes that were made?

Response

(*i*) After adding the ethanol, the solution is no longer 0.05 mol dm^{-3} in ammonium formate, nor does it contain 10% of ethanol.

(*ii*) The mobile phase was not filtered before use.

(*iii*) The mobile phase was not degassed before use.

Another common error in the preparation of this mobile phase is to use industrial grades of ethanol, which contain UV-absorbing impurities. The source of the ethanol should have been specified as well.

****

SAQ 9.2b

These questions refer to the properties of solvents that were listed in Fig. 9.2a.

(*i*) Considering a heptane/trichloromethane mobile phase, why would it be undesirable to replace the heptane by pentane, or the trichloromethane by tetrachloromethane?

(*ii*) Why is methanol or acetonitrile preferred to other water miscible solvents for the preparation of mobile phases for reverse phase chromatography?

(*iii*) What would be the major difficulty associated with the use of propanone in mobile phases?

Response

(*i*) Pentane is more expensive than hexane and is also much more volatile. Control of the mobile phase composition might be difficult.

Tetrachloromethane has a higher viscosity than trichloromethane (leading to a greater pressure drop across the column). It is also toxic.

(*ii*) Methanol is relatively cheap, has a low viscosity and low UV cut-off. Although more expensive and more toxic than methanol, acetonitrile has a lower UV cut-off. The others have higher viscosities, higher cut-offs and are more expensive than methanol. With ethanol, there are additional problems with Customs and Excise.

(*iii*) The major problem with propanone is the high UV cut-off (285 nm).

****

SAQ 9.2c

You are performing a separation using the following conditions:

Column: 5 μm C-18, 25 cm × 4.6 mm
Mobile phase: acetonitrile/phosphate buffer, pH 6.9, 75 : 25
Detector: UV absorption, 230 nm
Flow rate: 2 cm^3 min^{-1}
Temperature: ambient
Injection volume: 25 μl, using autosampler

The retention times that you get for the peaks vary from day to day; the table below shows some typical results.

| | | | | Retention time (min) | |
Day	Time	t_0	Peak 1	Peak 2	Peak 3
1	1000	1.25	2.17	12.12	17.84
	1300	1.26	2.19	12.17	17.82
	1600	1.24	2.18	12.15	17.85
2	0900	1.24	2.36	13.03	19.19
	1100	1.26	2.35	13.01	19.22
	1300	1.25	2.38	13.06	19.21

$\longrightarrow$

SAQ 9.2c (cont.)	Make a list of the more important experimental variables that are likely to affect retention in a reverse phase separation. Using the retention data above, is it possible to suggest which of these variables might be the cause of the problem?

Response

Four important variables that affect retention in a reverse phase separation are: flow rate, temperature, pH and mobile phase composition.

If the flow rate changes we would expect retention of all the peaks to change. Note that the retention time for an unretained solute, t_0, does not vary between days. The time for an unretained solute should be affected only by the flow rate, and not by the other variables under consideration, so a constant t_0 indicates that a flow rate change is probably not the cause of the problem.

Retention times in reverse phase chromatography typically decrease by 1–2% for a temperature increase of 1 °C. The retention time changes over the two days could be caused by a temperature difference of 5–10 °C between the two days. If the laboratory is subject to such temperature variations, we would be likely to get significant temperature changes within the day as well. Since the retention times are constant within the day, it seems likely that the retention time changes are not temperature-induced.

Changes in the mobile phase pH can cause large changes in retention time, but the magnitude and direction of these changes are seldom the same for all the peaks. The retention time of a neutral compound will not change with pH, but retention times of ionizable species may change in opposite directions, depending on the acidic or basic properties of the species and their pK values. Since the changes in retention time here are all of roughly the same magnitude and are all in the same direction, it is unlikely that a change in the mobile phase pH is causing the problem.

The discussion above indicates that the most likely factor that is causing the retention time changes in this case is change in the mobile phase composition. This could be caused by formulation errors as discussed in Section 9.2, by selective evaporation of one component of the mobile phase or, possibly, by the degassing procedure that was used.

The example was taken from J.W. Dolan, *LC–GC International* 1990, 3(12), 16–20, which contains other useful information on methods available for the isolation and prevention of such problems.

****

SAQ 9.3a

These questions refer to the extraction and separation of azines that was discussed in the previous Section:

(*i*) What difference between the azines and the PAHs is exploited to achieve their separation on the C-18 cartridge?

(*ii*) Why is HCl used when the cartridge is eluted?

(*iii*) If the PAHs had not been removed from the sample, how would you expect their retention times to compare with those of the corresponding azines (e.g. anthracene and acridine or naphthalene and quinoline)?

(*iv*) The chromatographic separation was done at pH 7.2. Can you think of any disadvantages of using a lower or a higher pH?

(*v*) Which detector could be used for the separation?

Response

(*i*) The azines are weakly basic compounds

(*ii*) The weak bases are protonated by the HCl. The protonated bases are highly polar species which elute rapidly from the cartridge, whereas the non-polar PAHs are retained.

(*iii*) The PAHs would be strongly retained by the non-polar stationary phase and their retention times would be much longer than those of the corresponding azines.

(*iv*) In acid buffers, the azines may be protonated. This would cause loss of efficiency and poor peak shape. The effect will diminish as the pH is increased, but at a pH above about 8 the lifetime of the column may be reduced, owing to dissolution of the stationary phase.

(*v*) As with PAHs, the conjugated structure of the azines suggests that fluorescence detection could be used. In fact, both fluorescence and UV absorption detection were used in this work.

****

SAQ 9.3b	These questions refer to the determination of dibasic acids in rhubarb, discussed in the previous section.
	(*i*) How is the precipitation of calcium oxalate prevented?
	(*ii*) What would be the order of elution of the two acids if the TBA was not present in the mobile phase?
	(*iii*) The TBA is an ion pairing reagent. What effect will it have on the retention of the two acids?
	(*iv*) What is the purpose of the C-18 extraction?

Response

(*i*) Calcium oxalate is soluble at pH 2, at which the oxalic acid is present predominantly as the monoanion.

(*ii*) At pH 2 the malic acid is present predominantly as the molecular form. The oxalic acid anion is much more polar and will elute from the column very rapidly.

(*iii*) The TBA will pair with the oxalic acid but not with the malic acid, which is not ionized. The effect of this is to increase the retention of the oxalic acid whilst the malic acid is unaffected. In fact, in this system the malic acid is eluted first.

(*iv*) Any non-polar constituents in the mixture will greatly increase the analysis time, as they will be strongly retained in this system and will take a long time to elute. The purpose of the C-18 extraction is to remove such constituents; for instance the red rhubarb colouring is retained on the extraction cartridge.

****

SAQ 9.3c	The following are all common practical problems when using HPLC. Suggest a probable cause of the problem and a remedy in each case.

(*i*) The pump motor is operating, but the pressure drop across the column suddenly falls to zero, and the flow of mobile phase stops.

(*ii*) Using a UV detector, there is a sudden upward step in the absorbance, or a series of large noise spikes on the chromatogram, or both effects at once.

$\longrightarrow$

(*iii*) Using a UV detector at 195 nm, the baseline is noisy, absorbance is high, and peak heights show poor reproducibility.

(*iv*) The pressure drop across the column increases until the cutout limit is reached, when the pump switches itself off.

Response

(*i*) The most likely cause of this is the presence of large air bubbles in the pump head (although a faulty check valve can sometimes produce the same symptoms). Degas the mobile phase, open the purge valve on the pump and pump mobile phase at a high flow rate for a while.

(*ii*) This is due to micro-bubbles in the detector cell (you may be able to see them emerging from the exit tube). Degas the mobile phase, disconnect the column and pass mobile phase through the detector cell at a very high flow rate.

(*iii*) At very low detection wavelengths, dissolved oxygen and other impurities will absorb strongly. If the mobile phase has been properly degassed, further purification of the mobile phase may be necessary. The reproducibility problem may well be due to small errors in setting the wavelength.

(*iv*) Assuming the cutout limit has been properly set, there is a blockage somewhere, usually in the inlet frit of the column. Sometimes the blockage can be removed by reversing the flow of mobile phase through the column for a while; alternatively, replace the inlet frit (some types of column are designed to try and prevent you doing this, in the hope that you will buy another one).

****

SAQ 9.3d	Comparing on-line and off-line methods of column switching, which of the following advantages and limitations do you think would apply to the on-line technique?

(*i*) Easier to carry out

(*ii*) Cheaper

(*iii*) Faster

(*iv*) Better reproducibility

(*v*) More chance of sample loss

Response

On-line techniques are easily automated but are more expensive as they require additional valves with associated switching equipment. Off-line methods are easier to carry out, but, because of the sample collection and re-injection steps, they are slower and tend to be less reproducible. With off-line techniques there is also more chance of sample loss due to adsorption or evaporation.

****

10. Some Additional Topics

10.1. SMALL BORE COLUMNS AND FAST LC

10.1.1. Small Bore Columns

The terms small bore, or microbore, refers to columns with diameters in the range 0.5–2 mm; the use of columns with diameter smaller than 0.5 mm is called micro LC. Microbore columns have, in principle, two advantages over conventional LC columns; one is that, because column performance depends on mobile phase velocity, they are operated at much lower flow rates, so there is a large reduction in solvent consumption and hence in operating costs (Fig. 2.4b). Microbore columns are operated at linear velocities similar to those used in larger diameter columns, corresponding to a much lower flow rate. If f (cm^3 min^{-1}) is the flow rate in a column with diameter d cm, and the mobile phase velocity is v cm min^{-1}, f and v are related by:

$$f = \frac{\pi d^2 v}{4}$$

(10.1a)

Also, if column 1, with diameter d_1, is operated at f_1, cm^3 min^{-1} and we want to operate column 2 which has a diameter d_2 at the same mobile phase velocity, then the flow rate that we need is given by:

$$f_2 = f_1 \times \frac{d_2{}^2}{d_1{}^2}$$

(10.1b)

∏   Use Eq. 10.1b to calculate the missing data in Fig. 10.1a. The correct figures are given at the end of the section. An example of the money that can be saved by operating at low flow rates is given in SAQ 10.1a.

Flow rate in 4.6 mm column, cm^3 min^{-1}	Flow rate in small bore colum at the same mobile phase velocity, μl min^{-1}	
	column diameter, mm	
	2	1
1		
2		
5	945	

Fig. 10.1a. *To be completed.*

The other advantage of microbore columns is that they produce small peak widths. If we compare a microbore and a conventional column and inject the same mass of solute, then for the microbore column the same mass of solute will be present in a smaller peak volume, i.e. the concentration of solute is higher. Since the detector responds to the concentration of solute, the microbore column will thus allow us to detect smaller quantities of material.

The use of microbore columns requires that the rest of the system has very low dispersion, and this is the main difficulty associated with their use. The volume of solute injected causes some dispersion, so for microbore columns this has to be small, usually less than 1 μl. There are several (expensive) small volume valve injectors that will do; the lowest volume available at the moment is 0.06 μl. It is also possible to use an external loop injector and switch it to the column for a short time only, so that the full loop volume is not delivered. Care has to be taken with the design of end fittings and connecting tubing so as to minimize dead volume, and the internal volume of the detector has to be reduced without causing loss of sensitivity. Electrochemical detectors are often used with small bore columns, as these are the easiest types to make with a small internal volume.

Columns of this type were first used as long ago as 1967, but at that time the influence of extra-column dispersion was not appreciated, so the columns were not used in properly designed apparatus. There was a renewal of interest in microbore techniques in the early to mid 1980s, when a great deal of development work took place, but since then the interest has waned. Presumably the driving forces of solvent savings and increased sensitivity for small samples have not been sufficient to convert users to microbore methods. It is also fair to say that the stringent dead volume requirements needed in the instrumentation have not been fully met by manufacturers, as they did not perceive the market to be there. There is, however, likely to be an increasing use of columns interfaced to spectroscopic detectors (especially MS), a continual demand for increased sensitivity, and higher costs, both for solvent and for solvent disposal. These factors may well result in a higher demand for microbore techniques in the future.

SAQ 10.1a

Suppose you are running a 4.6 mm HPLC column on a mixture of acetonitrile and water (80% by volume acetonitrile). The column runs continuously for 8 hours a day at a flow rate of 2 cm^3 min^{-1}, and your acetonitrile costs you £10.30 per dm^3.

(*i*) What is the cost of acetonitrile per year, assuming a year = 250 working days?

(*ii*) What is the mobile phase velocity (cm min^{-1}) through the column?

(*iii*) If you changed to a 1 mm column operated at the same velocity, what flow rate would you have to use?

(*iv*) What would the small bore column save you in acetonitrile costs?

SAQ 10.1a

10.1.2. Fast LC Separations

A typical HPLC column (25 cm × 4.6 mm, containing a 5 μm reverse phase packing) will have a plate number of 10 000–15 000. For many separations, this efficiency is far more than is needed, as often a plate number of 3000–5000 will give baseline resolution of all solutes. If this is the case, the use of a conventional column will waste both analysis time and solvent. Short columns (3.3 cm × 4.6 mm) packed with 3 μm bonded silica stationary phases have sufficient efficiency for many separations. They are commonly called 3 × 3 columns and have the same sort of advantages as are claimed for microbore columns, i.e. shorter analysis time, use of less solvent and higher mass sensitivity. They also require equipment with low dead volume, but this is not as serious a limitation as it is with microbore columns, and 3 × 3 columns can often be used satisfactorily in conventional instruments. However, the small particle size of the packing makes them susceptible to plugging problems.

Fig. 10.1b shows the separation of a mixture of six explosives, on a 3 × 3 column and on a conventional column. The same quantity of sample was used for each chromatogram. On the shorter column, a faster separation is obtained with a higher mass sensitivity.

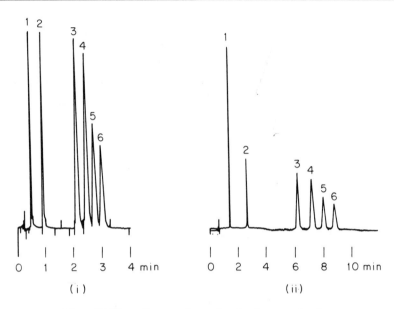

Fig. 10.1b.   *Separation of polynitro explosives*

Column: 5 μm C-8 bonded phase (*i*) 3.3 cm × 4.6 mm
                              (*ii*) 15 cm × 4.6 mm
Mobile phase: tetrahydrofuran/methanol/water 2 : 29 : 69
Flow rate: (*i*) 2 cm^3 min^{-1}, (*ii*) 3 cm^3 min^{-1}
Detector: UV absorption, 230 nm
Injection: 1 μl CH$_3$CN containing 100 ng of each solute
Sample: 1 HMX   2 RDX   3 2,4-DNT   4 Tetryl   5 TNT
               6 2,6- DNT

Very fast separations and very high efficiencies are possible with columns consisting of open or packed glass or silica capillaries, similar to those used in capillary GC. In the open columns the stationary phase can be formed by chemical modification of the internal surface of the tube. Such columns are still in the research stage, but their use is likely to increase in the future, as (like microbore columns) they are suitable for use with hyphenated techniques.

10.1.3. Summary

Small bore columns have a number of potential advantages in HPLC, especially for the in-line combination of HPLC with other techniques.

Learning Objectives

You should now be able to appreciate the advantages and limitations associated with the use of small bore columns.

10.2. SEPARATION OF ENANTIOMERS

10.2.1. Introduction

The separation of enantiomers (non-superimposable mirror image isomers) by HPLC is a technique that is becoming increasingly important, especially in the pharmaceutical industry. Many chiral drug compounds are manufactured and administered as racemic mixtures, and in some cases the enantiomers of such compounds differ substantially in their physiological effects. For instance, the dextrorotatory enantiomer (S)-(+)-methamphetamine (N-methyl-1-phenyl-2-aminopropane) is a controlled substance with a considerable history of drug abuse, whereas the (R)-(−) isomer is far less potent, and is an ingredient of proprietary decongestants. Similarly, the laevorotatory isomer of thalidomide is responsible for the teratogenic effects of the drug. The metabolism of enantiomers is frequently different, as enzymes responsible for biotransformations can distinguish between them, but the synthesis of enantiomers is often difficult, and the pharmaceutical industry has therefore become interested both in preparative chromatography as a means of producing optically pure drugs and in analytical separations to establish the optical composition of reaction products.

Several different approaches have been used for the separation of enantiomers. They can be separated as diastereoisomers on conventional columns, either after a suitable derivatization process or by the addition of chiral compounds to the mobile phase. Alternatively, a number of chiral stationary phases (CSPs) are available. These consist of chiral molecules bonded to microparticulate silica. They are considerably more expensive than conventional bonded silicas, and some types can be used only with a restricted range of mobile phases. The methods used for enantiomers are discussed briefly below.

10.2.2. Separation as Diastereoisomers

The principle of this method is, *deriving* that if we react a racemic mixture with a pure enantiomer of a chiral derivatizing reagent, the result will be the formation of diastereoisomers, which are compounds having more than one chiral centre which are not mirror images; these diastereoisomers have different physical properties and can be separated on conventional HPLC columns (e.g. reverse phase). The reagent used must be available in an optically pure form (and is therefore likely to be expensive) and must react with the racemic compound quickly, completely, and under mild conditions. For example, asymmetric drugs containing the amino group can be derivatized with acid halides to form diastereoisomic amides, with isocyanates to form ureas, or with isothiocyanates to form thioureas.

Diastereoisomeric complexes can also be formed by adding a chiral compound to the mobile phase. When this reacts with the analyte, racemic mixtures are resolved because of differences in the stabilities of the complexes, their solubility in the mobile phase or their interaction with the non-chiral stationary phase.

10.2.3. Chiral Stationary Phases

Pirkle CSPs are named after W.H. Pirkle (University of Illinois), who pioneered their development. These form a number of different selective interactions ($\pi-\pi$ bonding, hydrogen bonding, dipole or Van der Waals forces) between the CSP and individual enantiomers. They are probably the most widely used CSPs and there is now quite a lot of published work relating to them. The π-electron acceptor columns (used for separating π-electron donor enantiomers) use chiral dinitrobenzoyl derivatives of phenylglycine or leucine, bonded ionically or covalently to silica. The structure of the bonded phase is shown in Fig. 10.2a(i). They are available as (+), (−) or (±) forms and as analytical (5 μm) or preparative columns. The π-electron donor columns use a stationary phase consisting of naphthylalanine covalently bonded to 5 μm silica. These can be used to separate chiral amines, amino acids or alcohols that have been derivatized with a π-electron acceptor (e.g. dinitrobenzyol derivatives). The electron acceptor ionic CSPs have to be used with non-aqueous mobile phases of fairly low polarity (hydrocarbons modified with small amounts of alcohols). The covalent type CSPs are not restricted in this respect.

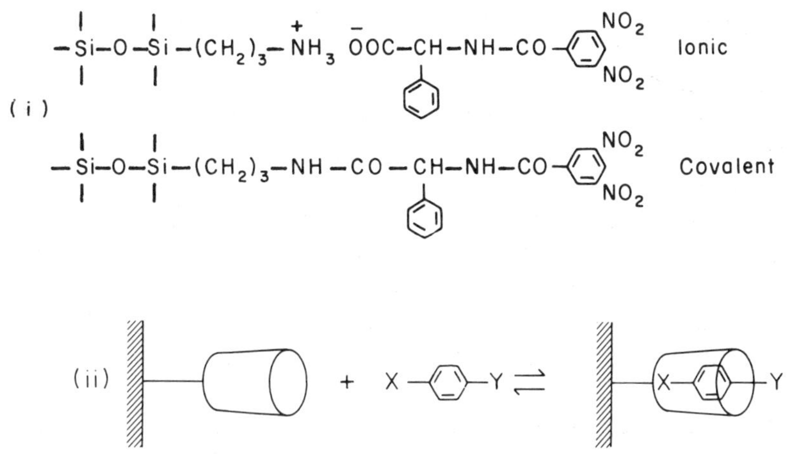

Fig. 10.2a. *Chiral stationary phases*

CSPs based on ligand exchange use a bonded chiral amino acid, e.g. *l*-proline complexed with Cu^{2+}, and a copper salt (about 10^{-3} dm^{-3}) modified with organic solvents as the mobile phase. They have been used to separate α-amino acids and other chiral compounds that can form chelate complexes with Cu^{2+}.

Inclusion complexes are entities comprising two or more molecules in which one of the molecules (the host) retains by physical forces a guest molecule. Cyclodextrins are typical host molecules. They are cyclic chiral carbohydrates, the CSPs being the α-, β- or, γ-cyclodextrins, which contain six, seven and eight glycopyranose units, respectively. They are bonded to silica via a 7 to 9 atom spacer between the support and the cyclodextrine molecule. The physical shape of the molecule is that of a truncated cone, with an internal cavity whose dimensions are determined by the number of glucose units. The structure of β-cyclodextrin is shown in Fig. 10.2b. The

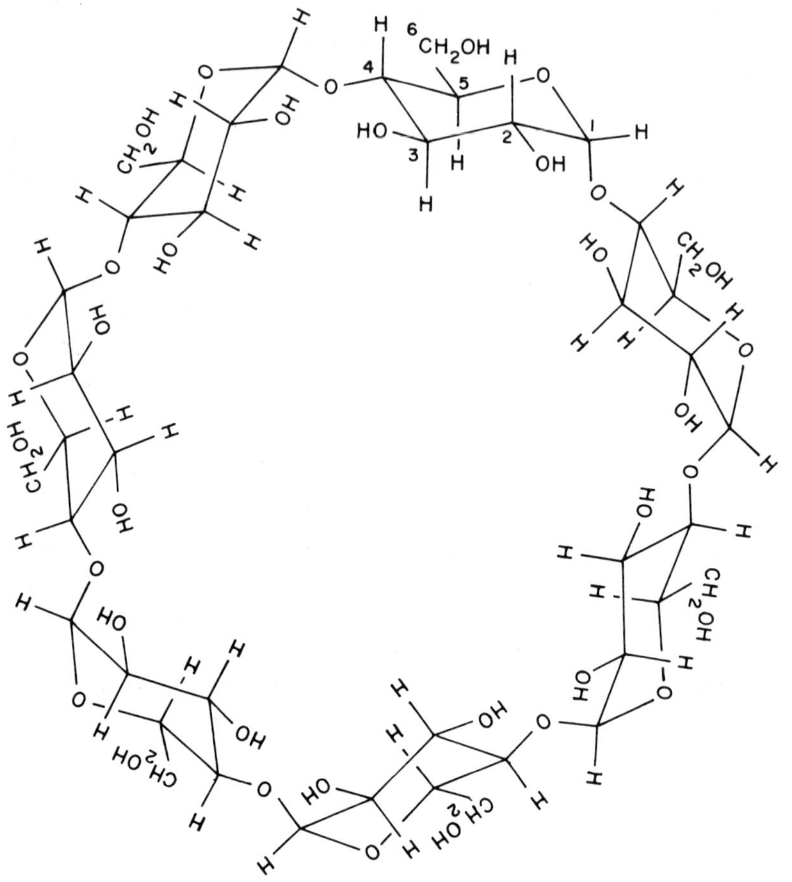

Fig. 10.2b. *Structure of β-cyclodextrin*

inner cavity is lined by the glycosidic oxygen bridges and is hydrophobic; the hydroxyl groups of adjacent units form hydrogen bonds which stabilize the shape of the molecule.

Solutes are separated by a mechanism that involves the formation of inclusion complexes, the strengths of which are governed mainly by the ability of the solute to fit into the cyclodextrin cavity. Thus the stationary phase can discriminate between solutes that differ only in geometry or spatial orientation, and has found use in the separation of a number of structural and positional isomers, as well as enantiomers. A simplified diagram of the inclusion process is shown in Fig. 10.2a(ii).

The mobile phases used with cyclodextrin CSPs are similar to those used in reverse phase chromatography, i.e. water or buffer solutions plus organic modifiers.

∏   The organic modifier competes with the solute molecules for inclusion in the cyclodextrin cavity. How would an increase in the concentration of modifier affect the retention time of a solute?

Increasing the concentration of organic modifier will decrease the interaction between the solute and the cavity and will thus lead to a shorter retention time. Similarly, a mobile phase modifier that forms a stronger complex will also reduce retention, e.g. ethanol forms a stronger inclusion complex than methanol.

Protein type CSPs are based on the interaction of acidic and basic drug compounds with albumin (for acidic drugs) and α_1-acid glycoprotein (for basic drugs). These proteins are polymers composed of naturally occurring chiral amino acid groups, and the interaction that occurs can often be stereospecific. These CSPs are used in the reverse phase mode, the mobile phase being a phosphate buffer modified with small amounts of propan-2-ol. Separation is achieved by careful adjustment of the pH, ionic strength and organic modifier concentration of the mobile phase.

10.2.4. Summary

The separation of enantiomers is an area of increasing importance. Enantiomers can be separated on conventional columns following a suitable derivatization, or by the addition of chiral compounds to the mobile phase, or by the use of chiral stationary phases.

Learning Objectives

You should now be able to:

• Recognize the areas in which the separation of chiral compounds is of interest;

• Describe briefly the different methods that are available for achieving such separations.

10.3. FLASH CHROMATOGRAPHY

This technique was first described by W.C. Still in 1978, and its development in the U.K. has been pioneered at May and Baker (now Rhône Poulenc). It occupies a position somewhere between the early gravity-driven LC methods and today's modern high pressure techniques. It is used mainly for the rapid isolation and purification of products from chemical reactions, in quantities ranging from 0.1–100 g.

The method uses a glass column, typically 30 cm long and 18–100 mm in diameter. This is packed *in situ* with 40–63 μm silica or other material (e.g. alumina or bonded silica) to a depth of about 15 cm. Dry packing is used for smaller diameter columns, slurry packing for larger ones. The sample to be separated is added in solution to the top of the column and is covered with a layer of acid-washed sand, to prevent disturbance when the column is filled with solvent. After adding solvent, the column is eluted with compressed air at 5–10 psi. Fig. 10.3 shows the experimental arrangement.

The eluent from the column can be monitored continuously with an in-line detector, or fractions can be collected. Method development and examination of separated fractions is usually done by TLC; in general, any compounds that can be separated by TLC can also be separated on a preparative scale by flash chromatography. The method is simple to set up and use, it is rapid (run times are typically 10–15 min) and the packings and solvents used are relatively cheap in comparison with HPLC grades.

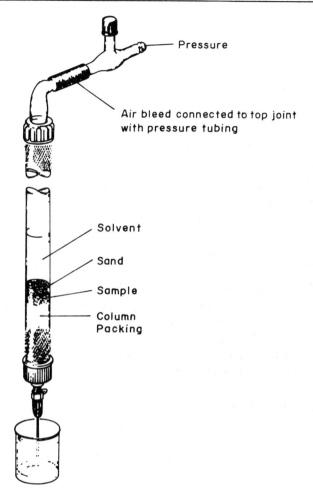

Fig. 10.3. *Experimental arrangement for flash chromatography. A safety screen should be used when the column is being eluted*

10.4. PREPARATIVE HPLC

HPLC separations may need to be scaled up for a number of reasons. We may need to separate a compound and isolate small amounts of it for confirmation of structure using other instrumental methods (e.g. IR, NMR or MS) or for elemental analysis. We may need larger quantities for further testing, or for the preparation of derivatives. Furthermore, we may want to use the technique as a method of commercial production, when large amounts of material will need to be separated.

Scaling up of HPLC separations always has to be done with care, because in general the efficiency of our system is likely to decrease if we increase the amount of sample applied to the column, the diameter of the packing particles, or the diameter of the column itself. Fig. 10.4a shows typical shapes of Van Deemter plots obtained (*i*) for a separation using analytical HPLC conditions, and (*ii*) for conditions that may obtain in a preparative separation. You can see from this that in the analytical separation we can shorten the analysis time by increasing flow rate without much loss of efficiency, but on the preparative scale it is important to operate the column at the flow optimum. It is helpful to perform method development with an analytical column packed with the same material that is to be used in the preparative scale separation.

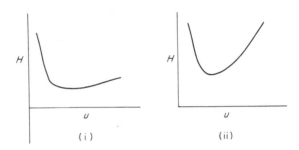

Fig. 10.4a.   *Van Deemter plots*

(*i*) Shape obtained for many analytical separations
(*ii*) Use of large column diameters or particle size causes considerable loss of efficiency at higher flow rates

Systems for semipreparative, preparative and large scale preparative separations are available commercially. For semipreparative work, the goal is usually to obtain small amounts of high purity material, so resolution is at a premium and semipreparative columns tend to be slightly larger versions of analytical columns, using the same packings. Preparative columns are typically 2–5 cm in diameter and 25 cm long with packings of 15–100 μm diameter. Columns for large scale work can be 20–30 cm in diameter and 60 cm long, using flow rates up to 1000 cm^3 min^{-1}. The commercial systems can be used isocratically or with gradients, and allow small scale development and preparative separation to be done using the same system. Large scale preparative work can become extremely expensive, both for the columns and packings and for solvent costs.

The following example outlines the use of preparative HPLC for the separation of benzodiazepam analogue enantiomers. As mentioned in Section 10.2, enantiomeric forms of drugs can have very different pharmacological effects; the first step in the development of many drugs often involves the testing of enantiomeric forms. The synthesized drug is likely to be racemic, in which case HPLC is used to isolate the two enantiomers. Benzodiazepam analogues are of interest because of their action as tranquillizers and sedatives.

The separation is first developed on an analytical scale, using a Pirkle type CSP column. Method development consists of (*i*) optimizing the binary mobile phase composition for maximum resolution and minimum analysis time, (*ii*) optimizing the flow rate, and (*iii*) performing a loading study to determine the maximum amount of sample that can be applied without unacceptable loss of efficiency. The chromatogram obtained using the maximum acceptable load is shown in Fig. 10.4b(i).

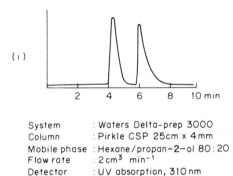

(i)

| | 2 | 4 | 6 | 8 | 10 min |

System   : Waters Delta-prep 3000
Column   : Pirkle CSP 25cm x 4mm
Mobile phase : Hexane/propan-2-ol 80:20
Flow rate   : 2cm³ min⁻¹
Detector   : UV absorption, 310nm
Sample mass : 3.2mg

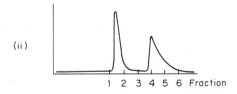

(ii)

| 1 2 3 4 5 6 Fraction |

Column   : Pirkle CSP 25cm x 21mm
Flow rate   : 50cm³ min⁻¹
Sample mass : 90mg

Fig. 10.4b. *Preparative separation of benzodiazepam enantiomers*

Preparative scale separation of the sample was done using the same CSP in a 25 cm × 21 mm column.

∏   Since this column should be operated at the same mobile phase velocity as the analytical column, what flow rate should be used?

Using Eq. 10.1b, the scale up factor is the ratio of the square of the column diameters:

$$f = 2.0 \times \frac{21^2}{4^2} = 55 \text{ cm}^3 \text{ min}^{-1}$$

The same scaling factor is used to determine the maximum sample loading on the larger column as 88 mg. The preparative chromatogram is illustrated in Fig. 10.4b(ii). Fractions were collected as shown, the first three fractions giving the (+) enantiomer with 99.4% purity and the fourth, fifth and sixth giving the (−) enantiomer with 99.6% purity.

10.4.1. Summary

The sample size in HPLC may be increased for a number of reasons; separations up to kilogram scale are now possible. Chromatographic efficiency generally decreases as the sample size is increased.

Learning Objectives

You should now be able to:

• Describe the situations in which preparative HPLC is useful;

• Appreciate some of the difficulties associated with the scaling up of analytical separations.

10.5. SFC

SFC, or supercritical fluid chromatography, uses as the mobile phase a supercritical fluid, that is, a gas above its critical pressure and temperature. Under these conditions the physical properties of the mobile phase are intermediate between those of a gas and a liquid. Supercritical fluid viscosities are relatively low compared with those of liquids, and diffusion rates in supercritical fluids are much higher than in liquids. These properties allow fast efficient separations to be done (as in GC) but using a mobile phase that has some solvating ability (as in HPLC). In addition, sensitive universal detectors like the flame ionization detector (FID) can be used.

The most common mobile phase in SFC is CO_2 (critical temperature 31.05 °C, critical pressure 72.9 atm). The low critical temperature allows the separation of thermally sensitive compounds, and also it is non-toxic, odourless and available in high purity. Although the solvent strength of a supercritical fluid can be increased by pressure programming, supercritical CO_2 is not very polar; under SFC conditions it has a solubility parameter of about 15 $(kPa)^{1/2}$. The range of fluids with higher polarities is very limited; they are either extremely toxic or, as is the case with the common polar solvents, have critical temperatures that are much too high—about the only practical possibility is ammonia. To circumvent this problem, the CO_2 is sometimes modified with polar solvents such as methanol, but under these conditions the system is incompatible with an FID.

SFC has been used with typical GC (packed and capillary) and with HPLC columns. It can be used for samples that would be unsuitable for GC analysis because of involatility and/or thermal instability and also for samples that would be difficult by HPLC because of detection problems. The technique is relatively easy to interface to other systems, e.g. MS or IR; because of this it is possible that the use of SFC will increase in the future.

10.6. LC–MS

There is only room here for a very brief mention of this complex and rapidly developing subject. The mass spectrometer separates and analyses ions in the gas phase. The analysis is fast, sensitive and amenable to examination of compounds which can be volatilized and/or ionized without degradation. In many ways the mass spectrometer represents the ideal HPLC detector in that it is highly sensitive, can be operated either as a universal or as a highly selective detector, and can in principle establish the identity of each eluted compound. No other detection system approaches MS in terms of this versatility, but this is only achieved at very high cost.

The earliest and most common means of sample ionization used is electron impact (EI), in which a chromatographically resolved analyte is vaporized and subjected to a beam of electrons. The beam is typically energetic enough not only to ionize the target molecule but also to break covalent bonds and fragment the molecule. The molecular fragmentation patterns produced are reproducible and represent a fingerprint of the molecule which can be recognized by the mass spectroscopist or computer-matched with standards.

The desire to work with labile compounds has led to the development of other methods of sample ionization including chemical ionization (CI), secondary ion mass spectrometry (SIMS), fast atom bombardment (FAB) and many others (refer to Ref 61 for details). Many instruments are now supplied with a variety of ion sources.

The major difficulty with the in-line combination of LC and MS is that the LC solutes are eluted in the presence of a vast excess of solvent and at atmospheric pressure. The solvent has to be removed, or the solute concentration has to be very greatly increased, before the eluent can be introduced into the MS, which operates at high vacuum. This is the function of LC–MS interfaces. Some of the more important devices are discussed briefly below.

(a) Moving Belt

This is the one of the earliest interfaces and is still available commercially. A metal belt passes below the HPLC column outlet and picks up a thin film of eluent. Solvent is removed by passing the belt through heated zones and a vacuum chamber before the residual material is flash vaporized within the vacuum system of the MS, close to the ion source. An advantage of the method is that both EI and CI can be used.

(b) DLI (Direct Liquid Introduction)

If solvent corresponding to typical LC flow rates (1 cm^3 min^{-1}) is vaporized, the result is about 20 times more vapour than the MS pumping system can handle. DLI methods either use a split at the column outlet, so that only a small fraction of the eluent is transferred to the MS, or involve use of micro-columns to reduce the eluent flow rates. Only chemical ionization can be used, and the range of solvents is limited. CI produces the molecular ion, with little or no fragmentation, so that the structural information obtainable is also limited.

(c) Thermospray

This is the most commonly used interface at present, and is compatible with typical HPLC flow rates. Aqueous mobile phases containing an electrolyte such as ammonium acetate are passed through an electrically heated stainless steel capillary situated in a heated ion source fitted with an auxiliary pumping line opposite the capillary. This produces an aerosol in which some of the droplets are charged. The size of the charged droplet is reduced as it travels through the heated ion source, consequently the electric field at the droplet surface increases until ions present in the liquid phase are ejected. Electrolyte ions (e.g. NH_4^+ can react with a sample molecule in the gas phase to generate analyte ions, which are sampled through a small aperture in the mass analyser. Again, only CI spectra are possible.

(d) Particle Beam Interfaces

These have recently been commercialized under the trade names of Thermabeam and MAGIC (monodispersive aerosol generation interface for combined LC–MS). In this method, a mobile phase aerosol is formed in a desolvation chamber, where the solvent is evaporated. The vapour is then passed through a two stage momentum separator in which the solvent vapour is pumped away, leaving only the sample particles to be transferred to the MS. The advantage of particle beam technology is that the hardware separates the processes of desolvation and ionization, and EI spectra can be produced.

10.6.1. Summary

SFC has potential advantages in areas of HPLC where detection is a problem. The combination of HPLC and MS provides a very sensitive and versatile separation method.

Learning Objectives

You should now be able to:

• Appreciate the advantages of using MS as an HPLC detector;

• Recognize the difficulties of in-line combination of the two methods;

• Outline the operation of some interfaces.

References

SMALL BORE COLUMNS

1. P. Kucera, *Microcolumn HPLC* Elsevier, 1984.

2. R.P.W. Scott, Ed., *Small Bore LC Columns, Their Properties and Uses*, Wiley, 1984.

3. C.F. Simpson, Ed., *Techniques in Liquid Chromatography*, Wiley, 1984, Chapter 4.

4. M. Verzele and C. Dewaele, *Journal of Chromatography* 1987, 395, 85–89.

SEPARATION OF CHIRAL COMPOUNDS

5. H.T. Karnes and M.A. Sarkar, *Pharmaceutical Research*, 1987, 4, 285.

6. J. Gal, *LC–GC* 1987, 5, 106.

FLASH AND PREPARATIVE CHROMATOGRAPHY

7. W.C. Still, M. Khan and A. Mitra, *Journal of Organic Chemistry* 1978, 43, 2923.

8. E. Lunt and C. Smith, *Laboratory Equipment Digest* Sept. 1989, 49.

9. B.D. Bidlingmeyer, Ed., *Preparative Liquid Chromatography*, Elsevier, 1987.

10. M. Verzele and C. Dewaele, *LC Magazine* 1985, 3, 22.

SFC

11. R.D. Smith, B.W. Wright and C.R. Youker, *Analytical Chemistry* 1988, 60, 1323A.

12. P.J. Schoenmakers and L.G.M. Uunk, *European Chromatography News* 1987, 1, 14.

LC–MS

13. R. Davis and M. Frearson, *ACOL Mass Spectrometry* Wiley, 1987.

14. T.R. Covey, E.D. Lee, A.P. Bruins and J.D. Henton, *Analytical Chemistry* 1986, 58(14), 1451A–1461A.

SAQ AND RESPONSE

SAQ 10.1a

Suppose you are running a 4.6 mm HPLC column on a mixture of acetonitrile and water (80% by volume acetonitrile). The column runs continuously for 8 hours a day at a flow rate of 2 cm^3 min^{-1}, and your acetonitrile costs you £10.30 per dm^3.

$\longrightarrow$

SAQ 10.1a
(cont.)

> (*i*) What is the cost of acetonitrile per year, assuming a year = 250 working days?
>
> (*ii*) What is the mobile phase velocity (cm min^{-1}) through the column?
>
> (*iii*) If you changed to a 1 mm column operated at the same velocity, what flow rate would you have to use?
>
> (*iv*) What would the small bore column save you in acetonitrile costs?

Response

(*i*) The column uses $2 \times 60 \times 8 = 960$ cm^3 of mobile phase per day or $960 \times 0.8 = 768$ cm^3 of acetonitrile per day or 192 dm^3 per year, which costs £1977.

(*ii*) $2 = \pi \times (0.46)^2 \times v/4$ $v = 12.03$ cm min^{-1}.

(*iii*) $f = \pi \times (0.1)^2 \times 12.03/4$ $f = 0.0945$ cm^3 min^{-1}.

(*iv*) Acetonitrile cost = £93.40, i.e. a saving of £1884 per year.

****

Units of Measurement

For historic reasons a number of different units of measurement have evolved to express quantity of the same thing. In the 1960s, many international scientific bodies recommended the standardisation of names and symbols and the adoption universally of a coherent set of units—the SI units (Système Internationale d'Unités)—based on the definition of five basic units: metre (m); kilogram (kg); second (s); ampere (A); mole (mol); and candela (cd).

The earlier literature references and some of the older text books, naturally use the older units. Even now many practicing scientists have not adopted the SI unit as their working unit. It is therefore necessary to know of the older units and be able to interconvert with SI units.

In this series of texts SI units are used as standard practice. However in areas of activity where their use has not become general practice, eg biologically based laboratories, the earlier defined units are used. This is explained in the study guide to each unit.

Table 1 shows some symbols and abbreviations commonly used in analytical chemistry; Table 5 is a glossary of abbreviations used in this particular text. Table 2 shows some of the alternative methods for expressing the values of physical quantities and the relationship to the value in SI units.

More details and definition of other units may be found in the *Manual of Symbols and Terminology for Physicochemical Quantities and Units*, Whiffen, 1979, Pergamon Press.

Table   1 *Symbols   and   Abbreviations   Commonly   used in Analytical Chemistry*

Å	Angstrom
$A_r(X)$	relative atomic mass of X
A	ampere
E or U	energy
G	Gibbs free energy (function)
H	enthalpy
	joule
K	kelvin ($273.15 + t$ °C)
K	equilibrium constant (with subscripts p, c, therm etc.)
K_a, K_b	acid and base ionisation constants
$M_r(X)$	relative molecular mass of X
N	newton (SI unit of force)
P	total pressure
s	standard deviation
T	temperature/K
V	volume
V	volt ($J\ A^{-1}\ s^{-1}$)
$a, a(A)$	activity, activity of A
c	concentration/ $mol\ dm^{-3}$
e	electron
g	gramme
i	current
s	second
t	temperature / °C
bp	boiling point
fp	freezing point
mp	melting point
$\approx$	approximately equal to
$<$	less than
$>$	greater than
e, $\exp(x)$	exponential of x
$\ln x$	natural logarithm of x; $\ln x = 2.303 \log x$
$\log x$	common logarithm of x to base 10

Table 2 *Alternative Methods of Expressing Various Physical Quantities*

1. **Mass (SI unit : kg)**

$$g = 10^{-3} \text{ kg}$$
$$mg = 10^{-3} \text{ g} = 10^{-6} \text{ kg}$$
$$\mu g = 10^{-6} \text{ g} = 10^{-9} \text{ kg}$$

2. **Length (SI unit : m)**

$$cm = 10^{-2} \text{ m}$$
$$\text{Å} = 10^{-10} \text{ m}$$
$$nm = 10^{-9} \text{ m} = 10\text{Å}$$
$$pm = 10^{-12} \text{ m} = 10^{-2} \text{ Å}$$

3. **Volume (SI unit : m^3)**

$$l = dm^3 = 10^{-3} \text{ m}^3$$
$$ml = cm^3 = 10^{-6} \text{ m}^3$$
$$\mu l = 10^{-3} \text{ cm}^3$$

4. **Concentration (SI units : mol m^{-3})**

$$M = \text{mol } l^{-1} = \text{mol dm}^{-3} = 10^3 \text{ mol m}^{-3}$$
$$\text{mg } l^{-1} = \mu g \text{ cm}^{-3} = ppm = 10^{-3} \text{ g dm}^{-3}$$
$$\mu g \text{ g}^{-1} = ppm = 10^{-6} \text{ g g}^{-1}$$
$$\text{ng cm}^{-3} = 10^{-6} \text{ g dm}^{-3}$$
$$\text{ng dm}^{-3} = \text{pg cm}^{-3}$$
$$\text{pg g}^{-1} = ppb = 10^{-12} \text{ g g}^{-1}$$
$$mg\% = 10^{-2} \text{ g dm}^{-3}$$
$$\mu g\% = 10^{-5} \text{ g dm}^{-3}$$

5. **Pressure (SI unit : $N \text{ m}^{-2} = \text{kg m}^{-1} \text{ s}^{-2}$)**

$$Pa = Nm^{-2}$$
$$atmos = 101\ 325 \text{ N m}^{-2}$$
$$bar = 10^5 \text{ N m}^{-2}$$
$$torr = mmHg = 133.322 \text{ N m}^{-2}$$

6. **Energy (SI unit : $J = \text{kg m}^2 \text{ s}^{-2}$)**

$$cal = 4.184 \text{ J}$$
$$erg = 10^{-7} \text{ J}$$
$$eV = 1.602 \times 10^{-19} \text{ J}$$

Table 3 *Prefixes for SI Units*

Fraction	Prefix	Symbol
10^{-1}	deci	d
10^{-2}	centi	c
10^{-3}	milli	m
10^{-6}	micro	μ
10^{-9}	nano	n
10^{-12}	pico	p
10^{-15}	femto	f
10^{-18}	atto	a

Multiple	Prefix	Symbol
10	deka	da
10^2	hecto	h
10^3	kilo	k
10^6	mega	M
10^9	giga	G
10^{12}	tera	T
10^{15}	peta	P
10^{18}	exa	E

Table 4 *Recommended Values of Physical Constants*

Physical constant	Symbol	Value
acceleration due to gravity	g	9.81 m s^{-2}
Avogadro constant	N_A	$6.022\ 05 \times 10^{2\delta3} \text{ mol}^{-1}$
Boltzmann constant	k	$1.380\ 66 \times 10^{-23} \text{ J K}^{-1}$
charge to mass ratio	e/m	$1.758\ 796 \times 10^{11} \text{ C kg}^{-1}$
electronic charge	e	$1.602\ 19 \times 10^{-19} \text{ C}$
Faraday constant	F	$9.648\ 46 \times 10^4 \text{ C mol}^{-1}$
gas constant	R	$8.314 \text{ J K}^{-1} \text{ mol}^{-1}$
'ice-point' temperature	T_{ice}	$273.150 \text{ K exactly}$
molar volume of ideal gas (stp)	V_m	$2.241\ 38 \times 10^{-2} \text{ m}^3 \text{ mol}^{-1}$
permittivity of a vacuum	ϵ_0	$8.854\ 188 \times 10^{-12} \text{ kg}^{-1} \text{ m}^{-3} \text{ s}^4 \text{ A}^2 \text{ (F m}^{-1})$
Planck constant	h	$6.626\ 2 \times 10^{-34} \text{ J s}$
standard atmosphere pressure	p	$101\ 325 \text{ N m}^{-2} \text{ exactly}$
atomic mass unit	m_u	$1.660\ 566 \times 10^{-27} \text{ kg}$
speed of light in a vacuum	c	$2.997\ 925 \times 10^8 \text{ ms}^{-1}$

Index

Absorbance ratios 84
Adsorption chromatography
 187–189
Amperometric detectors 91–92, 177
Applications data 250–251
Axial compression 67

Balanced density method 267
Beer–Lambert law 73

Capacity factor 17, 36, 210, 214,
 216, 272
Column extraction 288–290, 292
Column packings
 alumina 155–156
 bonded phase 2, 150–153, 155,
  162, 212
 carbon 155
 chiral 313–317
 microparticulate silica 2, 149,
  150, 193
 porous layer beads 149, 169, 286
 preparation of 150–152
 styrene–divinylbenzene resins
  154–155, 168, 191–192
Columns
 3 × 3 311–312
 axially and radially compressed
  67
 commercial 154
 dimensions 63
 efficiency 19
 fittings and threads 64–65

guard 285
Pinkerton 289
Pirkle 314, 321
reconditioning 286
scavenger 285
small bore 308–311
testing 272

Dead volume 5, 30–33, 70,
 282–283
Derivative preparation 55, 100–103,
 177
Derivative spectra 83
Detection limits 85, 238
Detectors
 calibration 230–246
 characteristics 69–71
 electrochemical 89–96
 fluorescence 87–88
 infrared 79
 mass spectrometric 323–326
 photodiode array 80–85
 refractive index 96–97
 sensitivity 70, 75
 UV-visible absorbance 73–86
Dispersion mechanisms 27–30

Eddy diffusion 27
Electrochemical detectors 89–96
End-capping 152, 162
Exclusion chromatography 189–198
External standard method 230–235,
 244, 246